AF750896

Scribbling on page 5

This book is to be returned on or before the last date stamped below.

UNIT No. 5

-5. JUN. 1976
12. JUN. 1976
26. JUN. 1976
22. JUL. 1976
-3. AUG. 1976
-1. SEP. 1976
12. NOV. 1976
21. JUN. 1977
-2. AUG. 1977
-5. SEP. 1977
-5. JAN. 1979
-9. FEB. 1979
-3. MAR. 1979
26. MAR. 1979
-6. APR. 1979
-3. SEP. 1979

North Walsham

-1. NOV. 1979
26. NOV. 1979
24. APR. 1980
20. JUN. 1980

H.Q. STORE

NORFOLK COUNTY LIBRARY

North Walsham

NORLINK 3 0129

ITEM 022 094 214

128.5

SHNEIDMAN Deaths of Man

DEATHS OF MAN

DEATHS OF MAN

Edwin S. Shneidman

millington

Copyright © 1973 by Edwin S. Shneidman
All rights reserved.
ISBN 0 86000 035 4

First British edition published in 1975 by
Millington Ltd.
109 Southampton Row
London WC1B 4HH

Reproduced and printed by photolithography and bound in
Great Britain at The Pitman Press, Bath.

~~990669~~

128·5

To
HENRY A. MURRAY
who taught me that life,
more vast than I had imagined,
was a wondrous process of discovery,
and that death,
while it might be explored,
can never be fully charted

“On life and death . . .”

Walking the deck with quick, side-lunging strides, Ahab commanded the t'gallant sails and royals to be set, and every stunsail spread. The best man in the ship must take the helm. Then, with every mast-head manned, the piled-up craft rolled down before the wind. The strange, upheaving, lifting tendency of the taffrail breeze filling the hollows of so many sails, made the buoyant, hovering deck to feel like air beneath the feet; while still she rushed along, as if two antagonistic influences were struggling in her—one to mount direct to heaven, the other to drive yawingly to some horizontal goal. And had you watched Ahab's face that night, you would have thought that in him also two different things were warring. While his one live leg made lively echoes along the deck, every stroke of his dead limb sounded like a coffin-tap. On life and death this old man walked.

Moby Dick, Chapter 51

CONTENTS

5 DIMENSIONS OF DEATH

FOREWORD

The author of this important and timely book has a continual acquaintance and an intimate knowledge of the spectacle of death. The distinction between the spectacle of other people's deaths and the event of one's own death is one of Professor Shneidman's key points. One's own death is an event in other people's experience, but it cannot be an experience of one's own; for death is, by definition, a permanent cessation of consciousness. This was also a key point for the Roman poet Lucretius in the first century B.C. In his poem *De Rerum Natura,* one of Lucretius's main concerns is to help his fellow human beings to overcome their dread of death. He is aware that one of the causes of distress is the virtual impossibility of thinking about the sequel to one's own death without falling into the illusion that the extinguished self is still in being and is still consciously observing the distressing consequences that obtrude themselves on the survivors.

Professor Shneidman discusses the spectacle of other people's deaths from a number of different angles. His concern is not merely professional; he is concerned humanely, both in easing a dying person's passage to death and in helping the survivors to contend with the psychological wound inflicted by bereavement (see Chapters 1 and 3 on death work and postvention—a useful new word). Professor Shneidman is also concerned as a scientist. His scientific field is psychology, and a substantial part of the present book is devoted to the "psychological autopsy." He rightly insists that both dying and killing have a psychological dimension. They are not simply physical events or acts, and therefore a death certificate that contains nothing beyond physical information is incomplete and may be misleading. Enterprising scientists gather their data from as wide and as highly diversified a field as they can survey. Professor Shneidman has found a psychological gold mine in the writings of Herman Melville.

The living generation's experience of the spectacle of death is peculiar. Today there are still millions of survivors of the two world wars, and of many local wars, who have both witnessed and inflicted man-made death by violence. But our bellicose age has also been an age of medical progress and consequently, in peace-time, and in war-time too, outside the war zones, most children now grow up with little or no first-hand experience of the spectacle of death—apart from the deaths of grandparents, and these cause little shock since it has become rare for grandparents to end their lives under the same roof as their children and grandchildren. However, the knowledge that genocide has become a possibility is now almost universal, and this among people of all ages. The author's chapter on "megadeath" deals with the effects of this knowledge on our attitudes and conduct.

In another chapter Professor Shneidman raises the question why, in our time, death is considered obscene. Feelings of obscenity are generated by the humiliating incongruity between the animality and the spirituality of human nature. Death, like excretion and sexual intercourse, is one of the cardinal features of our animality; and the indissoluble psychosomatic unity of human nature, which was taken for granted by primitive man, has been recognized once again by modern science.

Since our ancestors became human, we have been revolted by our animality, and we have dealt with this disquieting but unescapable aspect of our nature either by veiling it or by combating it. Modern man does both. He ignores death, yet at the same time treats death as his Enemy Number One. But it is difficult to box with death without personifying it; an enemy cannot be inanimate. However, death is not really a demon; it is an event, and an inevitable one. To try to wage war against this inevitable event is as futile as a dog's naïve reaction to stubbing his paw on a stone. He retaliates by biting the stone, at the risk of breaking his teeth after having bruised his paw. This canine reaction strikes human observers as being both comic and pathetic, and this should warn us not to fall into the "pathetic fallacy" in our emotional attitude toward death.

"Partial death" and survival after death in the "postself" are two fascinating topics of this book. Partial death means the loss, before physical death, of part of the self through bereavement or alienation or through the loss of some faculty. Survival in the postself means the pleasurable anticipation, while one is still alive, that after his death his life will produce posthumous effects in the world of men. Two such posthumous effects are progeny and fame. For survival through progeny, the Book of Genesis is the *locus classicus*. The anticipation of survival through fame has been expressed by the Roman poet Horace in the three words *Non omnis moriar* ("Not the whole of me is going to die").

The thirst for posthumous fame is shared alike by highly sophisticated writers and artists and thinkers and men of action, and the barbarian warlords immortalized by Homer. I could run on; but I will resist the temptation. The purpose of a foreword is to introduce a book, not to paraphrase it. I found the book most interesting, and I wish it every success.

ARNOLD TOYNBEE

The Royal Institute of International Affairs
Chatham House,
London
September 1971

ACKNOWLEDGMENTS

The wending of my way to this point in my life and my writing this book has (as the very word "wending" implies) been more fortuitous than planned. Serendipity, unusual opportunities, chance, and plain good luck have all played major roles in my life's fortunate professional course. For example: my career as a thanatologist began, in 1949, when I practically stumbled into a coroner's vault containing several hundred suicide notes. That event began an interest in the dark side of life which continues to this day.

Before I cite a number of specific acknowledgments, I need to make explicit my debt to two abiding forces that have sustained my labors.

One dominating aspect of my "wending" has been my twenty-year interest (obsession?) with the life and work of Herman Melville. (The very title of this book was, when it first occurred to me, an unconscious association to the ship *Rights of Man,* from which Melville's Billy Budd was kidnapped and pressed into the King's service.) The other major impact on my development has been the wisdom of my mentor and model, Dr. Henry A. Murray, to whom this book is dedicated. My love of Melville is second-hand: purloined directly from Dr. Murray's life-long study and classic writings on that genius. In light of all this, a more accurate title of this book might have been *The Deaths of Man: With Special Reference to Herman Melville and*

Henry A. Murray. As one who believes Dr. Murray's tenet that the study of lives is best pursued through the intimate understanding of the biographical and psychodynamic details of a relatively few uniquely interesting persons, I now share my interests in Melville and Murray with my readers unashamedly and with pleasure, hoping somehow that communication of the mysteries of life and death are, through the mediation of these two giants, facilitated between us.

The reader who wishes to become better acquainted with Herman Melville is encouraged to read that master—from *Typee* to *Billy Budd* with a long pull at *Moby Dick;* the reader who wishes to know more about Henry Murray can read Hiram Haydn's (1969) biographical sketch of him and two short autobiographical pieces (1959, 1967); and finally, the reader who wishes to know what Murray has written about Melville is referred to his comments on *Pierre* (1949), *Moby Dick* (1951), *Bartleby* (1966), and on Melville's general attitudes toward his own annihilation (1967). Dr. Murray is clinical Professor Emeritus at Harvard, embryologist, biochemist, physician, surgeon, psychoanalyst, clinical psychologist, and Melville scholar. He is eighty years old.

I am pleased to acknowledge the assistance of a number of willing helpers in the preparation of this book: members of the Psychology of Death class at Harvard in 1969, who provided the stimulus to write the book and whose thoughts and questionnaire responses are quoted liberally in it; Helen Lowrance, secretary, and Priscilla Jones, research librarian, at the Center for Advanced Study in the Behavioral Sciences, 1969-1970; Professor Howard Becker, for his help in straightening some of my thoughts about the format of this volume; Steven Brenner and Joseph Kaplon, students in the Death and Suicide class at UCLA in 1970, who helped with the preparation of Chapter 10; Vivian Schwartz, a member of the volunteer corps at the UCLA Neuropsychiatric Institute; and Karen Chase, a student at UCLA. Special thanks are due Mrs. Helene Collier, my secretary, who typed the manuscript. My wife, Jeanne, read the entire manuscript and made numerous suggestions for improvement, all of which I incorporated.

I am especially grateful to the late Rockwell Kent, who, shortly before he died in March 1971 at the age of eighty-eight, gave me permission to reproduce the haunting portrait of Captain Ahab which appears in the Random House edition of *Moby Dick* and which furnishes the frontispiece for this volume.

Appreciation is acknowledged to the following for permission to reproduce materials that appear in this volume:

Aldine-Atherton, Inc. for "Orientations toward Death" by Edwin S. Shneidman, from *The Study of Lives,* edited by Robert W. White.

Behavioral Publications, Inc. for *The Psychological Autopsy* by Avery D. Weisman and Robert Kastenbaum.

Communications/Research/Machines, Inc. for "You and Death" by Edwin S. Shneidman, from *Psychology Today,* June, 1971.

Dell Book Co. for "Reflections on the H Bomb" by Gunther Anders from *Man Alone: Alienation in Modern Society,* edited by Eric and Mary Josephson.

Farrar, Straus and Giroux for *Narcissus and Goldmund* and "Klein and Wagner" from *Klingsor's Last Summer* by Hermann Hesse.

Grove Press for "Han's Crime" by Naoya Shiga (translated by Ivan Morris) from *Modern Japanese Literature,* edited by Donald Keene.

Harcourt, Brace and World for *Young Radicals* by Kenneth Keniston.

Hogarth Press for "Thoughts for the Times on War and Death" from *Standard Edition of the Complete Works* by Sigmund Freud.

International Journal of Psychiatry for "Discussion of 'Orientations toward Death' " by Koji Sato, vol. 2, 1966.

Journal of Abnormal and Social Psychology for "What Should Psychologists Do about Psychoanalysis?" by Henry A. Murray, vol. 35, 1940.

Professor Abraham Kaplan of the University of Haifa for his unpublished manuscript, "The Self and Identity."

Kent State University Press for "The Deaths of Herman Melville" by Edwin S. Shneidman from *Melville and Hawthorne in the Berkshires,* edited by Howard P. Vincent.

Little, Brown and Co. for *Autobiography* (Vol. 2) by Bertrand Russell.

McGraw-Hill Book Co. for "The Role of the Social Scientist in the Medicolegal Certification of Death by Suicide" by Theodore J. Curphey from *The Cry for Help,* edited by Norman L. Farberow and Edwin S. Shneidman; for "Attitudes towards Death in Some Normal and Mentally Ill Populations" by Herman Feifel, from *The Meaning of Death,* edited by Herman Feifel; for "The Medical Definition of Death" by A. Keith Mant, from *Man's*

Concern with Death by Arnold Toynbee and others; and "Epilogue: The Relation between Life and Death, Living and Dying" by Arnold Toynbee, from *Man's Concern with Death* by Arnold Toynbee and others.

Macmillan Co. for *The Intelligent Individual and Society* by Percy Bridgman.

Pantheon Books for *The Politics of Experience* by R. D. Laing.

Penguin Books for *Dying* by John Hinton.

Prentice-Hall for *Passing On* by David Sudnow.

Psychiatry for "A Musical Idiot Savant: A Psychodynamic Study and Some Speculations on the Creative Process" by David S. Viscott, vol. 33, 1970.

Random House for *Life in Death: Survivors of Hiroshima* by Robert Jay Lifton.

Russell Sage Foundation for "Economic and Social Costs of Death" by Richard M. Bailey, "Innovations and Heroic Acts in Prolonging Life" by Robert J. Glaser, "Legal and Policy Issues in the Allocation of Death" by Bayless Manning, "What People Think about Death" by John W. Riley, Jr., and "Dying in a Public Hospital" by David Sudnow, from *The Dying Patient*, edited by Orville G. Brim, Jr., et al.

Schocken Books for *C. G. Jung and Hermann Hesse: A Record of Two Friendships by Miguel Serrano.*

Science House, Inc. for "Dead to the World: The Passions of Herman Melville" by Henry A. Murray, from *Essays in Self-Destruction*, edited by Edwin S. Shneidman.

Charles Scribner's Sons for *Twentieth Century Book of the Dead* by Gil Elliot.

Social Science and Medicine for "Life and Death: Dividing the Indivisible" by Richard Kalish, vol. 2, 1968; and "Psycho-Social Transitions: A Field for Study" by C. Murray Parkes, vol. 5, 1971.

U. S. Government Printing Office for *Mortality Trends in the United States: 1954-1963.*

West Publishing Co. for *Black's Law Dictionary* by Henry Campbell Black.

Yale University Press for *Letters of Herman Melville*, edited by Merrell R. Davis and William H. Gilman.

1

THE DYING PERSON —AND AFTER

1

DEATH WORK AND STAGES OF DYING

This book is about death. Being so, it concerns itself with the most mysterious, the most threatening, and the most tantalizing of all human phenomena. Death is destroyer and redeemer; it is the ultimate cruelty and the essence of release. Death is universally feared but, paradoxically, sometimes is actively sought. Although undeniably ubiquitous, death is incomprehensibly unique. Of all phenomena it is the most obvious and the least reportable; it encompasses the profoundest of man's perplexities and ambivalences. Over the ages death has been the source of fear, the focus of taboo, the occasion for poetry, the stimulus for philosophy—and remains the ultimate mystery in the life of each man.

"Death is the one thing you don't have to do. It will be done for you." So said philosopher Abraham Kaplan to me. But there is more to that statement than first meets the mind. Admittedly it implies that everyone dies, that no one can circumvent his own death. But it makes no attempt to explicate certain lesser truths: that persons die at various rates (from sudden to protracted), at various ages, and with various degrees of participation in their own death. In those special cases—nowadays rather frequent—in which a sensitive adult has been told (or knows) that he will die of a fatal disease within a relatively short time, he has the opportunity (and the chore thrust upon him) of preparing for his own death, that is, of doing his own "death work."

In the phrase "knows that he will die" the word "knows" has a special connotation that is critical in any discussion of death. To "know" that one is doomed to die soon is epistemologically different from "knowing" almost anything else about oneself or about the world. It is a kind of "knowing" infused not only with uncertainty but also with several layers of conscious and unconscious mental functioning. Of course, everyone "knows" he is going to die ("Life is a fatal disease"; "No one gets out of life alive"), but when one is seriously ill and has been informed that he is moribund, then his "knowing" about his imminent death is typically mixed with magic, hope, disbelief, and denial. After all, death has never happened to him before.

No one can ever really "know" that he is about to die. There is always the intermittent presence of denial. The recurrent denial of death during the dying process seems to be the manifestation of the therapeutic gyroscope of the psyche. (We must not forget that one of the principal functions of the personality is to protect itself against itself—against its own ravages, assaults, and threats.) And yet seemingly contradictorily—but not really contradictorily, because we are dealing with different levels of the human mind—people often do seem to "know" at some deep unconscious or primitive level when they are about to die and then, in a very special and identifiable way, they withdraw, perhaps to husband their last energies, to put themselves together and prepare for their end.

One implication of this view of "knowing" relates to the question "To tell or not to tell?" In the situation of a person with a fatal disease—as in many aspects of psychotherapy—there is almost anything that can be done badly and almost anything that can be done well. The issue of telling is thus not so much a question of whether or not, but of how, how much, how often, how euphemistically, how hopefully, and how far beyond what the patient (and various members of his family) already "knows" at that moment. Hinton (1957) and Weisman (1972) present excellent discussions of this complicated issue.

Death work imposes a two-fold burden: intrapsychic (preparing oneself for death) and interpersonal (preparing oneself in relation to loved ones and, simultaneously, preparing the loved ones to be survivors). The task is further compounded by pervasive emotional states: grief and anguish over the death to come; anger at one's impotence, at fate, and often, by unconscious extension, at key persons in one's life who are not dying, especially the young; and beneath these, a low-grade anxiety related to fear of pain, to loss of competence, to loneliness and abandonment, to fear of the unknown. But always, intermittently coming and going, is the fervent denial that death will really occur to him.

Dr. Elisabeth Kübler-Ross has delineated five psychological stages in the process of dying, based on her work with terminally ill patients in a Chicago hospital. Her deep concern with their welfare is evident; her book *On Death and Dying* (1969) is a clear manifesto of care for those who are benighted by the shadow of death. At the least, it stands as an antidote to some of the callous conventional hospital procedures surrounding the dying patient, described so well by Sudnow in *Passing On* (1967) and by Glaser and Strauss in *Awareness of Dying* (1965) and *Time for Dying* (1968).

Her book grew out of an interdisciplinary seminar at the University of Chicago, in the course of which dying patients were interviewed, or more accurately, were invited to speak of their fears and hopes, dreams and nightmares. The results of this unexampled open discussion were strikingly salutary for both the stressed dying person and the stressed hospital personnel. The interviews often have an evocative and haunting quality, resonating deep within one and stirring buried aspirations and fears. One cannot help being moved by the great human spirit in the voices of these dying fellow beings.

Kübler-Ross explicates five psychological stages of dying, or sets of reactions to one's awareness of imminent death. Categorized primarily "in terms of coping mechanisms at the time of a terminal illness," these stages are defined as: (1) denial and isolation ("No, not me; it can't be true!"); (2) anger—rage, envy, resentment ("Why me?"); (3) bargaining ("If you'll . . . then I'll . . ."); (4) depression ("What's the use?"); and (5) acceptance (the final rest before the long journey). According to her analysis, "the one thing that persists through all these stages [is] hope" ("I will not die"). One is reminded of Maurice Farber's *Theory of Suicide* (1968), in which self-destruction is seen as "a disease of hope," and of Kobler and Stotland's *The End of Hope* (1964), a case study of the death of a hospital.

Dr. Kübler-Ross does not tell us what percentages of the dying patients lived through each of these five stages or what the consequences were if any were cut off before they achieved the last stage. One key question is how one makes the transition from the negative affective states (which characterize the first four stages) to a state of acceptance. One might wish that she had extended her explorations of the nature of acceptance, as Henry Murray did so superbly in "Dead to the World: The Passions of Herman Melville" (1967). The last of Murray's five "psychic stages" in Melville's psychological development vis-à-vis death is "I accept my annihilation." But Murray then ponders the possible meanings of this acceptance, demonstrating that "acceptance" obviously has many dimensions:

> Did this last station of Melville's pilgrimage constitute a victory of the spirit, as some think? an ultimate reconciliation with God at the end of a lifelong quarrel? or was it a graceful acquiescence to the established morality and conventions of his world with Christian forgiveness toward those who had crushed him in their name? or a forthright willing of the obligatory? or was it an acknowledgment of defeat? a last-ditch surrender of his long quest for a new gospel of joy in this life? or was it a welcoming of death?

My own limited work has not led me to conclusions identical with those of Kübler-Ross. Indeed, while I have seen in dying persons isolation, envy, bargaining, depression, and acceptance, I do not believe that these are necessarily "stages" of the dying process, and I am not at all convinced that they are lived through in that order, or, for that matter, in any universal order. What I do see is a complicated clustering of intellectual and affective states, some fleeting, lasting for a moment or a day or a week, set, not unexpectedly, against the backdrop of that person's total personality, his "philosophy of life" (whether an essential optimism and gratitude to life or a pervasive pessimism and dour or suspicious orientation to life).

Philosophers—all but the twentieth-century analytic so-called philosophers—have traditionally taken life and death as their core topics. In relation to death and philosophy, the interested reader can turn to Choron's *Death and Western Thought* (1963) for a résumé of what major philosophers have thought about death. As to the tie between philosophic reflection and easing the burden of one's own death, I know of nothing more illuminating than Pepper's crisp and insightful essay, "Can a Philosophy Make One Philosophical?" (1967).

What of that nexus of emotions manifested by the dying person? Rather than the five definite stages discussed above, my experience leads me to posit a hive of affect, in which there is a constant coming and going. The emotional stages seem to include a constant interplay between disbelief and hope and, against these as background, a waxing and waning of anguish, terror, acquiescence and surrender, rage and envy, disinterest and ennui, pretense, taunting and daring and even yearning for death—all these in the context of bewilderment and pain.

One does not find a unidirectional movement through progressive stages so much as an alternation between acceptance and denial. Denial is a most interesting psychodynamic phenomenon. For a few consecutive days a dying person is capable of shocking a listener with the breathtaking candor of his profound acceptance of imminent death and the next day shock that listener with unrealistic talk of leaving the hospital and going on a trip. This interplay between acceptance and denial, between understanding what is happening and magically disbelieving its reality, may reflect a deeper dialogue of the total mind, involving different layers of conscious awareness of "knowing" and of needing not to know. Weisman's recent book, *On Dying and Denying* (1972), focuses on these complicated psychodynamics of dying, especially on the role of denial. One illustration of this ebb and flow of facing and denying the horror of death, especially the constant return of hope, can be seen in the account of a forty-eight-year-old dying woman reported later in this chapter.

The optimal working techniques for the clinical thanatologist (the professional or lay person, staff or volunteer) who deals directly with the death-laden aspects of the dying person have yet to be evolved. Admittedly, working with a dying person is a very special kind of intervention and would seem to require some special approaches and skills. As a beginning, one might say that the primary goals of such thanatological efforts are to help the dying person achieve a better death ("an appropriate death") and to help the survivors deal better with their loss, specifically to forestall morbidity or mortality. The interactions with the dying person are, understandably, different from almost any other therapeutic exchange, though they contain typical elements of rapport building, interview, conversation, psychotherapy (including interpretation), history taking, just plain talk, and communicative silences.

There are many ways in which a clinical thanatologist can find an appropriate occasion to speak with a dying person of feelings about death. Even dreams can be used to this end. (On a closely related topic, Litman, in 1964, wrote of the almost prophetic role of dreams in suicide). An example: A singularly beautiful and serene young woman, rapidly dying of metasticized cancer, had not talked about the possibility of her own death. One day, immediately after I came into her hospital room, she stated that she had had a dream the night before that she wanted to tell me. In her dream, she wandered down a street that she almost recognized but had never travelled, in a city that was vaguely familiar but one that she had never visited, and went to a house that seemed like home but clearly was not hers. In response to her ringing the bell, her mother answered the door, but that person was definitely no longer her mother. She then woke up.

"What does it all mean?" she asked me. In turn, I asked her what her mood had been during her dream. She said that it was one of pervasive sadness and nostalgia. She volunteered that perhaps the dream reflected her wish to be more independent of her mother, to lead her own life, living by her own standards. I asked her under what circumstances her mother might not be her mother. To this, she answered if her mother were dead, but she certainly did not wish for that; or, if it turned out that her mother were really not her biological parent, that she had been adopted, but that did not seem to make any sense. I went one step further: I asked if there might be any other circumstance. No, she said, she could not think of any. Her face then clouded and with some surprise she whispered, "If *I* were dead!" After she said this, her expression seemed to brighten. "Of course!" she said, and then asked, "Is that what I've been trying to keep from myself?" In the following few days she spoke of many details of her life, but intermittently, at a pace which she herself set, she talked about the realistic possibility of her death, her fear of dying, her concern for her loved ones, her deep regret at having to die, and the many joys she had had in her brief life. She said that she was glad that she and I could talk "honestly." Five days after her dream of death she died.

The thanatologist, if he examines his own mind, may very well discover that he is almost constantly aware of the expected "death trajectory" of his patient, and he governs the intensity of the sessions, their movement, the climaxes, the protective plateaus, and so on, over that projected time span—just as a skilled psychotherapist tries to control the intensity of the flow of material within any psychotherapeutic hour, trying not to leave the patient disturbed at the end of the session. The benign interventionist always needs to keep his hand on the rheostat governing the intensity of affect. The best results are achieved by turning affective intensity up gradually during the session and shading it down somewhat toward the end. So it is in the thanatological treatment of a dying person over the last months, weeks, days, hours, of his life.

In my belief, the transference and countertransference aspects of death work are unique, different in subtle ways from any other human exchange; both Hinton and Kübler-Ross mention this point.* For one thing, the situation itself, because of its obvious poignant quality and its time-limited feature, permits a depth of investment which in any other circumstances might border on the unseemly, yet in this setting is not only appropriate but perhaps even optimal. We can love a dying person, and permit a dying person to love us, in a meaningful way that is not possible in any other psychotherapeutic encounter.

It is difficult (and admittedly complicating) to speak meaningfully about Thanatos and its theoretical opposite, Eros. (Nor, parenthetically, do I view these concepts as anything other than figurative metaphoric polarities. I do not at all believe in a death instinct.) But considering the special nature of the transference and countertransference with dying persons, one might ask: Is it possible to fight death with love, or with loving care as a kind of substitute for love? Is it appropriate? Can it work? In some cases of dying persons I have worked with there was little question in my mind that Eros had helped to sustain them in their terminal periods and to lighten their dying days, that is, if by Eros we mean sympathy, support, and

*Hinton: "Often we came to know each other well—friendships grew fast in these circumstances and sometimes, it seemed difficult to believe we met only a few times"; and "The physician who had been questioned by this man [with cancer] said that he felt that he had probably been more perturbed by the conversation than the patient had" (1967, pp. 96, 134). Kübler-Ross "We find the same to be true among many patients who developed strong positive feelings towards us and were still able to pass through a stage of rage and anger often displaced onto others in the environment who reminded them of their own failing strength, vitality and functioning," (1972, p. 57).

concern, all the opposite of neglect.* One cannot always give the necessary transfusions of hope, but one should not fail to transmit one's honest feelings of interest, affection, and respect for the other's dignity—presaging one's own real sense of grief at the loss to come.

Working with dying persons seems to have some curious attractions and mystiques, some akin, I believe, to those associated with the very beginnings of life, the primal scene. In a sense the mysteries connected with sexuality are not as great as those lifelong uncertainties about the components of ideal love. In somewhat the same way, one can be continually intrigued, mystified, and even voyeuristic about working with individuals who are in the process of dying. As I see a dying person moving from day to day toward that ultimate moment, I hope that perhaps I can learn something about how dying is done, something about the arcane mysteries of the magic moment of transition from life to nonlife, something about the components of an ideal death, and even, if the gods are gracious, some guidelines that will teach me how to die well when my own turn comes.

Some practical words may be in order especially for the physician who sees a great number of dying persons in his practice—the oncologist and the hematologist primarily, but many others as well. A busy professional simply cannot take the time to deal intensively with each of his dying patients, and even if he had the time, he would not have the psychic reserves to do so. The physician would do well though to interact intensively with a few, perhaps only one dying person at a time, treating that person as a paradigm of all his dying patients. Further, a physician needs to take vacations from death. A gynecological oncologist, for example, should intersperse his practice with obstetrical cases, delivering babies as a balance for others of his patients who are dying of cancer of the uterus. And, as a further principle, a physician in oncological practice should not fail to seek out psychological or psychiatric consulation for his patients if they are significantly depressed or disturbed about dying and *for himself* if he senses that his own equanimity has been touched. This type of psychotherapeutic consultation might well be made a routine part of a physician's dealing with dying patients, lest he fall prey to the predictable consequences of the unusual psychological stress that comes from working constantly around and against death.

The remainder of this chapter presents verbatim excerpts from

*Arnold Toynbee, at eighty, says, "Love cannot save life from death; but it can fulfill life's purpose."

four videotaped interviews with a forty-eight-year-old woman who was dying of lymphosarcoma. (Some details have been changed, including her name. In this presentation we shall call her Lynn Cochran.) The first taped session was preceded by private sessions in her hospital room for a little over two months. I saw her almost every day, including some weekends, but excluding one week during which I was out of the country—an absence that had some distressing consequences. The four taped sessions are separated by two-week intervals.

Initially, I was asked to see her because of her behavior on the ward, especially her depression and withdrawal. Almost immediately after our sessions began, her doctor and nurses reported a dramatic change in her behavior: a reduction in her need for medication and a notable improvement in her outlook and demeanor. Of course, the principal concept which guided me—given the dire fact that she had a terminal illness—was Weisman's humane notion of an appropriate death. My goals were to try to help her die with a greater sense of completeness, of having resolved her relationships with her loved ones, of more inner tranquility; and less turmoil, reduced agitation, fewer loose ends, and diminished anger and anguish—in a word, to enable her to round out her life with a sense of accomplishment and peace.

A few words about the quality of the verbal interchanges between us: they are not, in any way, a model for clinical interviewing. Because of the special nature of the rapport—recall that I had seen her daily for some weeks before the first taped session—and perhaps because of the restraints imposed by the videotaping itself, what resulted differs in subtle but important ways from what went on in the private sessions in her hospital room.

FIRST SESSION

Patient: You couldn't move over or move my bed over a little bit, could you?

Shneidman: In what way?

P: I can't see you very well. I like to be able to see you.

S: Like this?

P: You're not on camera. [The camera was positioned behind a glass book case in an adjoining room.]

S: We don't have to worry about that.

P: You're a pretty good-looking guy. [An obvious transference remark: her concern at this moment seems to be almost entirely me. In a subsequent session she told me that seeing me was the most important event of her day. The special role of intense positive transference with a person known to be dying, especially when the thanatologist and the patient are of opposite sex, is a factor which obviously needs to be studied but which, based on my present intuitions ought to be elicited when possible. Hinton (1967, pp. 96, 134) mentions the difference in both the transference and countertransference phenomena between patients in ordinary psychotherapy and dying patients.]

S: The back of my head is my best profile. Now tell me what brought you into the hospital.

P: Well, actually a high fever, this time. When I came in they found that I had a low blood count and so they put me in the hospital.

S: Now, you have a diagnosis that you know. You've been told that.

P: Yes.

S: How do you understand that? What do you say to yourself is wrong with you?

P: I have lymphosarcoma.

S: What does that mean to you?

P: Well, as I understand it, it means that I have lymphs that are malignant and that apparently the lymphs are in my neck and my spleen and my groin and possibly other places, but I know those places that are most affected.

S: Malignant is a fairly heavy word that has all sorts of meanings.

P: Well, cancer.

S: Now that can be a frightening word to many people.

P: It is.

S: What does it mean to have cancer? What does it do to one's image of oneself and one's thoughts about life?

P: Well, when I first learned that I had it, it didn't mean too much. I wasn't affected for some time physically and I wasn't restricted.

S: At first it didn't seem real?

P: No, it didn't.

S: It was just one of those things.

P: That's right. Just like having a pimple. It didn't bother me any more than that. I had no pain.... [Usually diagnoses do not have the reality impact for the patient that they do for the relatives. What is real for the patient is pain, weakness, vertigo, disfigurement, and the burden of being confined to a hospital bed. Her relatives were initially much more disturbed by the diagnosis than she was. She had to reassure them.]

S: As I remember it, you first noticed this in yourself as a little lump in your groin.

P: Right.

S: Then you didn't make much of that. You thought maybe it was a little hernia or something like that.

P: That's what I thought. I didn't have a regular doctor, as I was getting ready to make a move back here where I had lived for a great many years, and I wanted to see my regular doctor here. And there was no indication it was anything urgent. There was no pain. It wasn't getting bigger. It was just a little less than half a marble. You could feel it down there. So when I got back I went to my regular physician. He examined me on the table and said that I had an infection below my waist. And he didn't think it was anything to worry about. But he did want to see me again in two weeks. He said, now if it does start to change size or anything, call me. And of course, I never dreamed of cancer in myself, you know. It's been all around me. I've had friends with it, relatives with it, but as for me, I never thought once of having anything like that.

S: To thee and thee, but not to me.

P: Not to me. So, I went back. I believe I had three visits after that at two-week intervals. And he said, it's not going to amount to anything. He said, don't worry about it. But he said if it does change size or anything, give me a call. So I went with it for about a year. Then I believe that I had the same thing happen on the other side of my groin, so I called my doctor immediately and went in. He was a private practitioner but near him was a clinic where they took blood tests and all kinds of tests, and he sent me there for various tests. And my son felt that I should be having all these tests, you know, to get them over with. Apparently he had the idea before I did that it might be something serious. [Was this a missed opportunity to have made an early diagnosis of the cancer?]

S: You had described what was wrong with you to your son?

P: Oh, sure, he's the one that's responsible for me coming here.

S: How old is your son?

P: Well, Bobby is twenty-eight.

S: Yes, so that he had some sense about this.

P: Well, he had more than I did . . .*

S: You have talked about hope.

P: I'm not always hopeful. I have my bad moments.

S: Let's talk about those moments, too.

P: You have your bad moments, but you try to avoid them and get over them and not dwell on them.

S: But there's a legitimate place for them. What do the dark moments look like?

P: Oh, I don't know, except when I was very ill, they looked pretty bleak.

S: A couple of days ago you were . . .

P: I was pretty bleak

S: Yes, you were really in the dumps. Please talk about that, won't you?

P: Well, I had for some reason, you know, I had this colostomy operation, which was completely unexpected.

S: They just came in and . . .

P: That's right. On Friday night they came in and on Saturday it was done. I had no idea there was anything like that I had to face. It was a surprise and a shock that I had to cope with that then . . . . [She had an anal thrombosis, which developed into gangrene and cellulitis. The colostomy was necessary. Her doctor, without going into a detailed technical explanation, simply told her that she had to have it. Of subsequent psychological interest was the way in which she adjusted to the colostomy; what under most circumstances would have been a major event in a person's life was, in the context of her dying, simply a great annoyance that, paradoxically, had helped to keep her alive.]

P: I feel as though I had this enemy.

S: What about this enemy that's in you?

P: I don't really know much about him. You know, he's kind of an elusive thing that I've never seen him.

S: You've seen some of his effects.

*In light of some later tragic developments, these passing references to her son (and those in the next session) take on an especially poignant meaning—see Chapter 12 for a detailed account.

P: Yes, I've seen some of his effects. The fact that I don't heal. I understand that that's responsible.

S: You've seen it in your loss of weight, your loss of energy.

P: Oh, yes. It's frightening.

S: Yes, of course. How frightened are you of all this?

P: I don't . . . I'm not really frightened. Maybe I have been frightened occasionally. You know, moments when I was on my own here.

S: What do you fear most, dear? What's the greatest fear?

P: That I won't be able to live a normal-type life.

S: Do you have fears beyond that?

P: Well, I probably do.

S: Can you verbalize them?

P: I'm not sure I can.

S: Now, the other night you mentioned a truly dirty word.

P: Yeah, we've said a few dirty words.

S: Do you want to come back to those?

P: Well, certainly, whatever you'd like to ask me.

S: Well, what was that really dirty word we talked about?

P: Death [Open talk of death with a dying person is a complicated thing but, on balance, seems to have generally salutary effects when the topic is—as it always should be—initiated by the patient.]

S: What are your thoughts and fears in relation to death?

P: Mine are probably similar to yours if you've ever had a very serious illness or maybe close to an accident. I once just missed a very serious accident driving across country, and I think it's a similar thing to that. You don't know what death is . . .

S: What do you believe about what doctors should do? Should they tell the patient?

P: I think so. For me, at least.

S: You'd prefer to know.

P: Oh, absolutely. . . .

S: Most people would prefer not to know when. How do you feel about that?

P: Well, my doctors say they don't know when.

S: No one really knows.

P: So how would I know?

S: No one knows. But it's an interesting question, if someone could know . . .

P: Oh, I don't think that would be a good thing. . . . [The seeming contradiction between her wishing to know what is wrong with her and her wishing not to know when she will die makes, of course, psychological sense.]

SECOND SESSION

S: How does time move when you're in the hospital?

P: It drags. You wake up in the morning, and generally I sleep pretty close to breakfast time. And then you look forward to having your bath and your dressings changed. When your dressings are changed, you're clean. Then you have a little freedom, you feel like lying there, watch television, or go to sleep, or whatever you want to do. And then, of course, during the day it's just to look forward to the tray and the dressing. And somebody like you stops in and sees me, and then that's the highlight, you know.

S: The big event of the day.

P: Yes, when I see you.

S: Who else has come to see you recently?

P: Well, you know I have other friends who would come, but I don't particularly want them to come. All I'd have to do is pick up the phone and say, come on over here. But I just really haven't wanted that.

S: Why is that?

P: Well, because, you know, I haven't been well, and it takes energy just to talk, really. And there's some people that I know, I have one friend—I'd just love to see her, and we'd have something to say and I'd enjoy as long as she'd stay. But most people I see exhaust me after a short while.

S: When your friends come, what would they and you talk about?

P: Well, I don't know. Our relationship is certainly not on the same level than that of my relatives or even a good friend, you know. I wouldn't talk to them about most things I say to you.

S: Like what, dear? Can you mention some of them?

P: Well, things we talk about, you know, death and illness and what I have to look forward to, you know. . . .

S: You haven't mentioned your son, Bobby. When did you see him last?

P: It's been two nights ago. Sometimes he comes four or five times a week.

S: As often as that?

P: He'd come whenever I picked up the phone and called him. But there are times when I've said, Bobby, I don't want to see you tonight. He's a night owl.

S: I guess that's why I miss seeing him.

P: Well, I'd really rather see him in the afternoon and he knows it, but he's just not a day guy; he's a night guy. . . .

S: You were talking about some infections.

P: Well, you know, in the last couple weeks I've had some infections come up. Like now, I have an infection in my bloodstream. And I had a spasm in my back here that kept me from getting up. I had been getting up and walking a little bit and this put me back into bed.

S: Those must have been scary things to happen.

P: Oh, you know they are. I've never heard anyone say they did, really . . . Well, we really don't know what dying is, we know we're out of it, you know, like sleep. And we don't know anybody who has died and can tell us what it's all about.

S: That's really frightening.

P: So I thought that if you have a fear, that's probably what it stems from. Aren't you almost always afraid of things like that?

S: What do you think will happen? [A rather straightforward remark—implying the probability of imminent death.]

P: It depends. I imagine it would depend on the state of my health at the time. You know, if I was in the hospital or, I mean, I could be fairly . . . I think I could be . . . the place where I might even be up and out. I mean it could happen shortly after that, I don't know. Or it could be like this, and if it were like this, I doubt if I'd—I doubt very seriously if I'd know. The doctors would probably drug me and I probably wouldn't know what was happening. [The denial of her terminal state—"I might even be up and out"—coupled with the wish for a gentle death.]

S: You'd slip into unconsciousness?

P: Probably. Don't you think?

S: I think so. . . .

P: I don't know why I have this.

S: How is it you got it? How do you explain that to yourself?

P: I don't think I can answer that.

S: Have you thought about it at all?

P: No, not really. I've accepted that fact that I have it and that's a change you have to accept.

S: It is just one of those chance things?

P: Possibly, or possibly it was something that could have been coming on for some years, who knows? They say stress has a lot to do with it, you know—cancer patients, they found that a lot of them have been under stress and it could play a role in maybe activating this illness. But now I don't know if that's it at all, but apparently there are a lot of things that could activate the illness.

S: Do you think of things you should have done or shouldn't have done?

P: No.

S: That you've been a bad person in any way?

P: No.

S: Do you see it in any way as punishment?

P: No. What would you do? You wouldn't know what to do. You wouldn't say five years from now I'm going to have a malignancy so I'm not going to do this, you know. You can't do that. You don't know. I never dreamed I'd be ill like this. It just never dawned on me.

S: You imply in what you say that there's been a good deal of stress in your life, that you've had a stressful life. Do you feel that way?

P: Yes, I've had quite a bit of stress. You know, I've been in business for twenty years and it's, you know, a lot of pressure.

S: You've been on your own a great deal, too.

P: Oh, yes. For a long time. Possibly ever since I lost both my parents on the same day.

S: I recall your saying that you were twelve years old then. That's rather young to be an orphan. . . .

P: I love to talk to you.

S: You do?

P: Certainly.

S: I think our doing this videotape every couple of weeks is a good idea, don't you?

P: Oh, if it will help the doctors, medical students, or nurses, whoever watches the film, to see a different angle.

S: Lynn, you haven't seen yourself on the videotape, have you?
P: No.
S: Would you want to?
P: Oh, yes, I want to.
S: You're so beautiful on camera.
P: How about you?

THIRD SESSION

S: How are you feeling today?
P: I really feel better. I feel better now than I have felt, you know, I mean at the mental end. I sort of look forward to seeing you, and that's maybe part of it.
S: Now, part of what happened in the last week was that I was out of the country.
P: Yes.
S: Our communication wasn't very good—namely, none of the postcards that I wrote you were delivered till after midweek, and then all at once. So that we were really out of touch for four or five days.
P: Well, you saw me on Friday and then you were away until Monday a week after that.
S: And you didn't get those postcards until Thursday.
P: No.
S: Although I mailed the first one on Saturday. I hear that on Wednesday you had a kind of flurry of feeling and reaction and so on.
P: I haven't felt too well.
S: I wondered if you'd talk about that. What happened last Wednesday evening. [I had mailed her a postcard each day, including one the day before I left. The foreign (Mexican) airmail was extremely efficient, but the cards were held up for five days in the hospital mailroom. They were all delivered on Thursday; on Wednesday night she had hysterics.]
P: Well, since you left on Friday a week ago I had a terrible depressive state of mind for about four or five days and I couldn't seem to get out of it. You know, it was just almost overwhelming. . . .
S: How did you get over it?

P: Well, I just told myself that I had to get out of it.

S: It was a sort of dangerous state.

P: Yes, it was.

S: What would have happened, do you think, if you'd have stayed in that state?

P: I don't know what would have happened, I don't think I cared at the time.

S: Have you any idea what might have happened?

P: No, but I certainly wasn't getting better if I stayed in that mental state I was in.

S: Are you saying you could have died in that state?

P: Possibly.

S: When was that, Wednesday night?

P: Well, that was while you were away, and I think it was on Wednesday night. I had a talk with a doctor and I told him exactly how I felt and I think that helped me some, and it was after I talked with him I decided I had to get out of that state of mind.

S: You were tearful, you were crying.

P: Oh, yes, yes.

S: Sobbing?

P: Oh, yes.

S: Really.

P: Very emotional.

S: I've not seen you like that, have I?

P: Well, not possibly as bad as I was.

S: I've seen you cry.

P: Yes. I just didn't really care.

S: You didn't care? Does that mean you didn't care about anything, whether you lived or died?

P: Really.

S: Really. You're back to caring now, aren't you?

P: Yes.

S: Yes. You're all dressed up, and you're coiffured, and I think you're wearing earrings.

P: Yes, they came from Mexico City. A very dear friend gave them to me.

S: I'm sure of that.

P: And it was a terrible feeling. You see, I've been here now a long time, I know there are patients who have been in the hospital longer than I've been in the hospital, but it's to me almost endless, you know, and I just began to wonder if I'd ever get out. Well, those days that we're talking about, that's all I could think about.

S: That's all you could think about, wondering whether or not you'd ever get out. How do you feel about that today?

P: I'm hopeful.

S: That hope has come back.

P: Yes. It becomes more difficult, Dr. Shneidman. You see, in my case, I've become weaker and I've lost weight, you know, quite a bit of weight and it becomes—seems to me harder to keep up your state of mind where it's, your mental outlook, it's at least acceptable, you know. . . . When am I going to be able to function? At least somewhat, so that I have some kind of life outside of bed?

S: When we first met, you had some rather definite dates in mind. What has happened to those?

P: Fell by the wayside.

S: Yes, so now you're sort of making different bargains with fate.

P: That's right. Right now I'm not making any bargains. I'm just going to try to keep my mental state better and do anything I can do to help me to get out of here. . . . [The effects of my absence and of the break in communication—through the inadvertent nondelivery of the postcards—are clear, as is the waxing and waning of hope. "To get out of here" is a euphemism for renewed or continued life, a reversal of the downhill course toward death.]

FOURTH SESSION

P: Right now I just have no energy or anything.

S: Now, today, dear, you seem lower on energy than on any of our previous sessions. How are your spirits? [It was evident from her gaunt appearance that she was ill and weak, almost moribund.]

P: Everything's bad when I feel that way. I'm afraid I'm not going to be too enlightening today, but do ask me what you'd like to and I'll try to answer.

S: All right, dear. . . .

P: I'll answer, I won't talk any more.

S: You're so touchingly cooperative.

P: Well, I want to cooperate. I want these sessions to mean something to the people that see them.

S: I can assure you they do . . . Are there any questions we've talked about, any issues we've talked about over the last couple of months that especially come to your mind at this time?

P: Only my getting out of the hospital.

S: The question of what the future will bring?

P: Yes.

S: How does the future loom in your mind?

P: Well, I just wonder when I'll get out of here. But I'll see you tomorrow.

S: Then, I'll see you tomorrow, of course.

P: Okay, Doctor.

Postscript: I saw her the next day and the day following that. On each visit we spoke only briefly. On the second day, painfully, she asked if she had been useful, if our filmed sessions would be helpful to medical students, to doctors, to hospital personnel, and to others. She hoped that *she* had been helpful. I assured her that she was marvelous in every way and said that I would see her the next day, to which she replied, "Good-bye, darling." I never saw her again. She died in her sleep at 2 A.M. the following morning. When the ward nurse telephoned me at home around 7 A.M. to tell me of her death, I remember that my first reflection was that death levels some people more than others. This remarkable person grew in her death and in that process taught us all.

Added to the tragedy of her death was a still more shocking sequel. The day of her death I telephoned both her son and another relative, neither of whom answered. I concluded that they had gone to her home state for her funeral. A week later I telephoned again, only to be told by the relative that Lynn's son had died the very day after her death. He had been fatally burned in a fire in his apartment. The mode of his death—whether accident or suicide—was not immediately clear. A sober consideration of the various implications of his death provides a set of interwoven themes throughout this volume. His death illustrates the dyadic (two-person) aspect of death (Chapter 5); it dramatically underscores the need for working with the principal survivor of the dying person (Chapter 3); in its cryptic nature, it exemplifies the concept of an "equivocal" death, in which the correct mode of death is initially unclear (Chapter 9); it raises questions about the role of the decedent in his own demise—what we call a "subintentioned" death (Chapter 8); and it provided the sad occasion for conducting an intensive psychological study subsequent to his death, called a "psychological autopsy," in which an equivocal death is investigated with the aim of arriving at an accurate certification of its mode (Chapter 12). The sequelae and "remnants" of death—the autopsy, disposition of the corpse, funeral, certification of death, processing of the will—all demonstrate that for the survivor the death of a loved one is not only an ending, but always, inexorably, a beginning as well. It is a pain that most of us learn to bear for many decades, but for some few (like Lynn's son) may be tolerated for only a few tortured hours.

2

AN APPROPRIATE DEATH

Every once in a while some gifted student of human nature enunciates an especially felicitous concept: Freud's idea of dreams as wish fulfillment, Kohler's concept of insight, Lawrence Frank's notion of projective techniques, Jung's archetypes, Tolman's cognitive maps, Maslow's peak experiences or Murray's need achievement. Weisman's (1972) concept of appropriate death belongs, it seems to me, in such company. It is illuminating, elevating, just right, and on the humanitarian side of things. In essence, the basic notion—resting on the assumption that some deaths are better than others, just as some lives are better than others—is that the beneficent thanatologist (physician, psychologist, nurse, or clergyman) should do what he can, when interacting with a dying person, to help him achieve a "good" death.

Weisman defines an appropriate death as "one in which there is reduction of conflict, compatibility with the ego ideal, continuity of significant relationships, and consummation of prevailing wishes. In short, an appropriate death is one which a person might choose for himself had he an option. It is not merely conclusive; it is consummatory."

A lovely and brilliant young woman who was dying of scleroderma said to me, "I badly don't want to die, but most of all, I don't want to die badly." She had just been told by her physician that she had only about a year to live. He told me, after she died seven days later, that he had really believed that she had had about a month to live, and he had been trying both to assuage and to alert her. Properly alerted—but also made anxious and sad—she used those last few days to say final farewells to her husband and her mother. Her words to them were words of love, concern, and courage. Psychologically she died rather well, with loose ends tied up, relationships cemented, her mind relatively at rest. By talking openly about her death with those she loved most, she made it possible for them to come closer to her rather than forcing them to pull away into painful silence or meaningless small talk.

We must distinguish among a timely death, a significant death, a wished-for (preferred) death, and a good death. The English physician John Hinton (1967) sees a timely death as one "entirely appropriate [in old age] . . . when a person has completed his span of life, his powers wane, and the eventual increasing decline indicates that it is time for the individual to depart this life." On the other hand, Dr. Kurt Eissler (1955) believes that "death always comes both too early and too late—too early because the ego has rarely realized all its potentialities, and too late because individual life has been a detour leading finally to what it had been at the beginning: nothingness."

Professor Koji Sato (1967) of Kyoto University, commenting on varying orientations toward death, has given a striking illustration of a timely death:

> Even Zen masters have their individuality, and their ways of leaving this world are different from one another. Master Gempo Yamamoto's way of passing away is specially interesting. After the Master reached his ninetieth birthday, he was thinking of closing the curtain of his puppet show in this world. He began fasting a number of times. At one o'clock on the night of June 3, the Master requested a cup of wine from his attendant monk, took it and showed his enjoyment. After ten minutes he said briefly: "I am going out on a journey, prepare my clothing," and lay down and passed away at 1:15 A.M. in his ninety-sixth year. Someone said that it was suicide, but it is quite sure it was not a usual suicide. The fruit was ripe enough. Unusually

ripe fruits fall down unintentionally, but this fruit fell down intentionally, being ripe enough.

At the other end of life, a young college student said it this way: "When I can say 'I've lived my life,' it is time to die with a smile on my face, being fulfilled in life and death. I cannot forsee the specific situation, because I do not know the path my life will take."

With a special introspective intensity, Abraham Maslow (1970) has related his feelings about a meaningful (timely) death following a serious heart attack that had occurred soon after he had completed writing a book that was important to him:

> I had really spent myself. This was the best I could do, and here was not only a good time to die but I was even willing to die. It was what David Levy called the "completion of the act." It was like a good ending, a good close. I think actors and dramatists have that sense of the right moment for a good ending, with a phenomenological sense of good completion—that there is nothing more that you could add.

Many men wish their death—the act itself—to stand for something; they fear an insignificant death. Sacrifice, ritual suicide (hara kiri, seppuku), death for a principle, death at the apogee of a life—these are often preferred to pointless or anticlimatic deaths. Stephen Crane echoed this wish for a significant death in *The Red Badge of Courage* (Chapter 21):

> In this last length of journey the men began to show strange emotions. They hurried with nervous fear. Some who had been dark and unfaltering in the grimmest moments now could not conceal an anxiety that made them frantic. It was perhaps that they dreaded to be killed in insignificant ways after the times for proper military deaths had passed. Or, perhaps, they thought it would be too ironical to get killed at the portals of safety. With backwards looks of perturbation, they hastened.

And is it not just this feeling that gives Baumer's death in Remarque's *All Quiet on the Western Front* its special ironic poignancy?

When asked about the kind of death he would prefer, the average man says that he would wish to die without protracted suffering and pain, rather quickly and without demeanment and humiliation. Deaths by accident, by heart attack, and during sleep are especially preferred. If the quick and painless death can occur in the performance of an act of heroism or at some peak in life, so much the better. "The only way to die that makes any sense is on your bike or fighting for your brothers," said a member of a New York City motorcycle club (*The New York Times*, November 24, 1969).

When we speak of a good death, we imply that it is appropriate not only for the decedent, but also for the principal survivors—a death they can "live with." The death is somehow consistent with the decedent's living image. While death itself is the ultimate violation, the nature of the death has not violated the image of the person who has died. In James Agee's evocative masterpiece *A Death in the Family*, Andrew tells Mary that Jay, her husband, died utterly unfraid. He was killed instantly; he was only briefly aware of the danger to him, and then it ended. Andrew says it this way: "Danger made him every inch of the man he was. And the next instant it was all over." His expression at the moment of death was "startled, resolute, and mad as hell. Not one trace of fear or pain." Mary is comforted to hear that her husband died without suffering and without weakness. "Very appropriate," her mother says. And having grasped the appropriateness of Jay's death, she chooses an equally appropriate epitaph for his gravestone: "In his strength."

Herman Hesse, who thought and wrote much about the meaning of death, has a particularly telling passage in one of the works of his early middle age, "Klein and Wagner" (in *Klingsor's Last Summer*):

> There was a spark floating in the darkness. He clung to it with all the ardor of his racked soul. It was a thought: useless to kill himself, to kill himself now; no point to exterminating himself, tearing himself limb from limb—it was useless. But it was good and redeeming to suffer, to ferment to ripeness amid tears and tortures, to be forged to completion amid blows and pangs. Then you had earned the right to die, and then dying was good, beautiful, and meaningful, the greatest blessing in the world, more blissful than any night of love: burned out and utterly resigned to fall back into the womb, to be extinguished, redeemed, reborn. Such a death, such a ripe and good,

> noble death, alone had meaning; it alone was salvation, it alone was homecoming. Longing cried in his heart. Where, where was the narrow, difficult path, where was the gateway? He was ready; with every quiver of his exhausted, agitated body, of his anguished mind, he yearned for it.

The psychologically noteworthy aspects of this passage are the terms Hesse chooses to characterize an appropriate death: to ferment to ripeness; to be forged to completion; burned out and utterly resigned; good; noble; meaningful; salvation. To die at the right time and in the right way—those are the hallmarks of a sterling death.

We all recognize that after certain peaks in life's range, there are anticlimatic plateaus where life hangs heavy and time stands still. Shakespeare reminds us in *Julius Caesar*. "There is a tide in the affairs of men. . . ." An appropriate death comes when one is ready for it. Avery Weisman (1966) quotes a doctor's comforting words to a dying patient: "I promise you that you will not die until you're prepared to die."

An appropriate death is appropriate to the individual's time of life, to his style of life, to his situation in life, to his mission (aspiration, goals, wishes) in life; and it is appropriate to the significant others in his life. Obviously, what is appropriate differs from person to person: one man's nemesis is another man's passion. Appropriateness has many dimensions, relating, at the least, to the state of one's health, competence, energy, prowess, zeal, hope, pain, and investment in his postself (see Chapter 4).

Suicide and homicide present some especially interesting issues in this context. (I am speaking here of homicide as the term is ordinarily understood—totally unexpected assassination, for example—rather than victim-precipitated homicide or homicide with any participative elements.) Although Weisman (1966) cautions us that the idea of appropriate death may be entertained as "merely a poetic metaphor or a sentimental afterthought when someone has died," it is nonetheless tempting to ruminate about the appropriateness of death from a historic point of view.

Two examples, each with its own enigmatic implications: To those of us who admired both the work and the life of Ernest Hemingway, would it not have been more "appropriate"—certainly more syntonic with his life style and his image in our minds—if he had been killed in that airplane crash on Kilimanjaro in Africa, rather than by blasting off his own head in Idaho after suffering the pain and indignities of electroshock treatments? We are saying, in effect, that the old man deserved better of fate—or are we really talking about our own petty preferences rather than his wishes and needs? And the very mention of the assassination of President Kennedy is apt to produce a concatenation of diverse and warring feelings: one may wish him alive and vibrant today, and rank his assassination with Lincoln's as a national tragedy, while at the same time reflecting that there could have been no more Greek-god death for him, no more dramatic end to a meteoric career, no more ennobling climax. What more "appropriate" death could he have had?

The professional literature dealing with the dying—what to tell, how much to tell, how to tell, when to tell, whom to tell—is somewhat confusing if not contradictory, though some good general guidelines are available (Brim 1970, Hinton 1967, Kübler-Ross 1969, Weisman 1966.) Weisman and Hackett (1967) believe that "planning and implementating psychotherapeutic interventions in patients who are certain to die may be facilitated if these criteria"—reduction of conflict, attention to ego ideal, continuity of important relationships, and consummation of wishes—"are used as guides, together with the psychodynamic formulation of the patient's concerns."

In general, one's goal in working with dying persons, their relatives, and their physicians and nurses—each of whom may bear psychological burdens distinctly his own in content and magnitude—is to be as candid and open about death as one can, consistent with the individual's capacity, at that moment, for "knowing." (Total honesty in almost any situation can be destructive.) *How* he knows is the core issue; though he is capable of "knowing" the fact that he is dying, he will modulate it in a number of ways, consciously and subconsciously, with wide variations in insight, repression, denial, understanding, terror, and equanimity. Where death is concerned, no one knows the facts for sure; those details are known with certainty only after they have occurred—too late for the protagonist.

But people working with a dying person can help by feeding him information at the rate at which he desires it—though there will always be a degree of ambivalence in his desire to know. One of the chief functions of the personality, to protect itself from its own inner workings, is not abandoned toward the end of one's life; it is maintained in force, if not actually increased. A dying person will not permit himself to hear more than he is prepared to digest at that moment. He very rarely "knows" more about his condition than is good for him.

A somewhat extended vignette may clarify some of these points. A fifty-year-old Greek-born man entered a large hospital with advanced leukemia. I was asked by his hematologist to see him because he seemed depressed. My initial impression of him is still vivid: he was a virile, handsome, straightforward man, an innocent and honest man, in the way that a rural or unsophisticated man can be innocent. Until a few weeks before he had worked as a construction laborer. He had been in a hospital only once before in his life, for treatment of a cut on his hand. His present complaint was that he felt weak. He had already been told that he had leukemia, but he could not remember the name of the disease and simply asked, "Is it a serious sickness?" His main concern seemed to be that his wife should be spared worry; he did not want her to know that anything serious was wrong with him.

During our third interview he said that on the previous day a young "hippie doctor" had talked to him and had asked him, "Are you afraid to die?" He asked me what I thought of a man who would ask such a question. I said that it was obviously a bad question, but added that perhaps someday, with his permission, he and I could discuss that very question. He repeated that it was a terrible thing to ask a man who had, after all, just come into the hospital and was sick. He said that the question had bothered him so much that he had quarreled with his wife on the telephone.

When I saw him the next day he was visibly shaking, and he was very much troubled about this tremor. "Why do I shake?" he asked. His concern about his tremulousness persisted over the next several days, and I used this concern as part of my therapy with him. When he said, "Oh, I know why I shake. It's the sickness no?" I said, "No, I didn't think so." "Ah," he said, "of course it's not the sickness–it's the medicine no?" I said no to that too, quietly conveying some conviction. If it wasn't the disease and wasn't the medicine, what could it be, then? I asked him if he had ever shaken that way before. He said never. I asked, "Never, not even as a little boy?" Well, yes, of course, he said, if I wanted to count that. As a youngster, in the old country, he had shaken once, when he had gone into a haunted house in a small village. Had he been frightened, scared? Never, he said; he wasn't scared of anything in this world. That was my cue to say, "*This* world?" Well, he admitted, anyone would shake if he was scared. What was he scared of now? Well, nothing in this world, but "I have to admit," he said almost casually, "I'm scared of death."

His ability to admit his fear presaged, it seemed to me, a number of important changes. After that he was able to talk realistically to his wife, even to speak about such painful details as his insurance, his funeral, their children, and his hopes for their education and future. With close friends and relatives, in his voluble style and with beautiful and appropriate emotion he was able to imply and even express his farewells. He took stock of his life and counted his blessings. As an act of courage and as a statement of his faith, he decided against the involvement of a priest in his last days. His loved ones could begin to feel their grief and convey to him their appreciation, love, and sense of what his death would mean to them. Positive sentiments were exchanged. There was much talk, even laughter, around his bed. Everything now seemed honest and aboveboard. A few weeks later he was dead. I think he died more appropriately than he would have done if the topic of death has been suppressed, unspoken, taboo.

One might question the value of these efforts to help the dying person to acknowledge his coming death and deal with it realistically. Inasmuch as he is doomed to die soon in any event, what is gained? I believe there are considerable gains, over and above their obvious sentimental appeal. For one thing, they improve the general quality of medical and nursing care because they force the medical personnel to regard the patient as the living person he is, rather than merely a "case"; they elevate ward morale within a hospital; they may lead to essential mental health care for the bereaved; they reaffirm the key role of such verities as candor, honesty, and dignity. They are practical simply because they are humane. They ennoble both the dying person and the one who attempts to sustain him.

3

POSTVENTION AND SURVIVOR-VICTIM

A person's death is not only an ending; it is also a beginning–for the survivors. Indeed, in the case of suicide the largest public health problem is neither the prevention of suicide (about 25,000 suicides are reported each year in the United States but the actual number is much higher, probably twice the reported rate) nor the management of suicide attempts (about eight times the number of reported committed suicides), but the alleviation of the effects of stress in the survivor-victims of suicidal deaths, whose lives are forever changed and who, over a period of years, number in the millions.

This is the process I have called "postvention": those appropriate and helpful acts that come *after* the dire event itself (1968, 1971). The reader will recognize prevention, intervention, and postvention as roughly synonymous with the traditional health concepts of primary, secondary, and tertiary prevention, or with concepts like immunization, treatment, and rehabilitation. Lindemann (1944) has referred to "preventive intervention in a four-year-old child whose father committed suicide"; it would be simpler to speak of postvention.

Postvention, then, consists of those activities that serve to reduce the aftereffects of a traumatic event in the lives of the survivors. Its purpose is to help survivors live longer, more productively, and less stressfully than they are likely to do otherwise.

It is obvious that some deaths are more stigmatizing or traumatic than others: death by murder, by the negligence of oneself or some other person, or by suicide. Survivor-victims of such deaths are invaded by an unhealthy complex of disturbing emotions: shame, guilt, hatred, perplexity. They are obsessed with thoughts about the death, seeking reasons, casting blame, and often punishing themselves.

The recent investigations of widows by Dr. C. M. Parkes (1972) are most illuminating. The principal finding of his studies is that independent of her age, a woman who has lost a husband recently is more likely to die (from alcoholism, malnutrition, or a variety of disorders related to neglect of self, disregard of a prescribed medical regimen or commonsense precautions, or even a seemingly unconscious boredom with life) or to be physically ill or emotionally disturbed than nonwidowed women. The findings seem to imply that grief is itself a dire process, almost akin to a disease, and that there are subtle factors at work that can take a heavy toll unless they are treated and controlled.

These striking results had been intuitively known long before they were empirically demonstrated. The efforts of Erich Lindemann (1944), Gerald Caplan (1964), and Phyllis R. Silverman (1969) to aid survivors of "heavy deaths" were postventions based on the premise of heightened risk in bereaved persons. Lindemann's work (which led to his formulations of acute grief and crisis intervention) began with his treatment of the survivors of the tragic Coconut Grove night-club fire in Boston in 1942, in which 499 people died. Phyllis Silverman's projects, under the direction of Gerald Caplan, have centered around a widow-to-widow program. These efforts bear obvious similarities with the programs of "befriending" practiced by the Samaritans, an organization founded by the Reverend Chad Varah (1966) and most active in Great Britain.

On the basis of work with parents of adolescent (fifteen- to nineteen-year-old) suicides in Philadelphia, Herzog (1968) has enumerated three psychological stages of postventive care: (1) resuscitation: working with the initial shock of grief in the first twenty-four hours; (2) rehabilitation: consultations with family members from the first month to about the sixth month; and (3) renewal: the healthy tapering off of the mourning process, from six months on.

A case can be made for viewing the sudden death of a loved one as a disaster and, using the verbal bridge provided by that concept, learning from the professional literature on conventionally recognized disasters—sudden, unexpected events, such as earthquakes and large-scale explosions, that cause a large number of deaths and have widespread effects. Martha Wolfenstein (1957) has described a "disaster syndrome": a "combination of emotional dullness, unresponsiveness to outer stimulation and inhibition of activity. The individual who has just undergone disaster is apt to suffer from at least a transitory sense of worthlessness; his usual capacity for self-love becomes impaired."

A similar psychological contraction is seen in the initial shock reaction to catastrophic news—death, failure, disclosure, disgrace, the keenest personal loss. Studies of a disastrous ship sinking by P. Friedman and L. Lum (1958) and of the effects of a tornado by A. F. Wallace (1956) both describe an initial psychic shock followed by motor retardation, flattening of affect, somnolence, amnesia, and suggestibility. There is marked increase in dependency needs with regressive behavior and traumatic loss of feelings of identity, and, overall, a kind of "affective anesthesia." There is an unhealthy docility, a cowed and subdued reaction. One is reminded of Lifton's (1967) description of "psychic closing off" and "psychic numbing" among the *Hibakusha*, the survivors of the atomic bomb dropped on Hiroshima:

> Very quickly—sometimes within minutes or even seconds—*Hibakusha* began to undergo a process of "psychic closing off"; that is, they simply ceased to feel. They had a clear sense of what was happening around them, but their emotional reactions were unconsciously turned off. Others' immersion in larger responsibilities was accompanied by a greater form of closing off which might be termed "psychic numbing." Psychic closing off could be transient or it could extend itself, over days and even months, into more lasting psychic numbing. In the latter cases it merged with feelings of depression and despair. . . . In response to this general pattern of disintegration, *Hibakusha* did not seem to develop clearcut psychiatric syndromes. To describe the emotional state that did develop they frequently used a term which means a state of despondency, abstraction or emptiness, and may be translated as "state of

> collapse" or "vacuum state." Also relevant is a related state . . . a listlessness, withdrawn countenance, "expression of wanting nothing more," or what has been called in other contexts "the thousand-mile stare." Conditions like the "vacuum state" or "thousand-mile stare" may be thought of as apathy but are also profound expressions of despair: a form of severe and prolonged psychic numbing in which the survivor's responses to his environment are reduced to a minimum—often to those necessary to keep him alive—and in which he feels divested of the capacity either to wish or will . . . related forms of psychic numbing occur in people undergoing acute grief reactions as survivors of the deaths of family members—here vividly conveyed in a psychiatric commentary by Erich Lindemann (1944):
>
> "A typical report is this, 'I go through all the motions of living. I look after my children. I do my errands. I go to social functions, but it is like being in a play; it doesn't really concern me. I can't have any warm feelings. If I would have any feelings at all I would be angry with everybody.' . . . The absence of emotional display in this patient's face and actions was quite striking. Her face had a mask-like appearance, her movements were formal, stilted, robot-like, without the fine play of emotional expression."

All this sounds remarkably like Henry Murray's description of partial death, those "psychic states characterized by a marked diminution or near-cessation of affect involving both hemispheres of concern, the inner and the outer world."

Postventive efforts are not limited to this initial stage of shock, but are more often directed to the longer haul, the day-to-day living with grief over a year or more following the first shock of loss. Postvention is in the honored tradition of holding a wake or sitting *shiva;* but it means more. Typically it extends over months during that critical first year, and it shares many of the characteristics of psychotherapy: talk, abreaction, interpretation, reassurance, direction, and even gentle confrontation. It provides an arena for the expression of guarded emotions, especially such negative affective states as anger, shame, and guilt. It puts a measure of stability into the grieving person's life and provides an interpersonal relationship with the therapist which can be genuine, in that honest feelings need not be suppressed or dissembled.

An example may be useful: Late one afternoon a beautiful nineteen-year-old girl was stabbed to death by an apparent would-be rapist in a government building. Within an hour her parents were hit between the eyes with the news (and this is the way these matters are usually handled) by two rather young, well-meaning, but inexperienced policemen. The victim was the couple's only child. Their immediate reactions were shock, disbelief, overwhelming grief, and mounting rage, most of it directed at the agency where the murder had occurred.

A few days later, right after the funeral, they were in the office of a high official who was attempting to tender his condolences when the mother said in an anguished tone: "There is nothing you can do!" To which, with good presence of mind, he answered that while it was true the girl could not be brought back to life, there was something that could be done. Whether he knew the term or not, it was postvention that he had in mind.

I began seeing the parents, usually together, sometimes separately. The principal psychological feature was the mother's anger. I permitted her to voice her grief and to vent her rage (sometimes even at me), while I retained the role of the voice of reason: empathizing with their state, recognizing the legitimacy of their feelings when I could, but not agreeing when in good conscience I could not agree. I felt that I was truly their friend, and I believed they felt so too.

A few months after the brutal murder, the mother developed serious symptoms that required major surgery, from which she made a good recovery. I had insisted that each of them see a physician for a physical examination. Although the mother had had similar difficulty some years before, the situation raises the intriguing (and unanswerable) question in my mind whether or not that organic flurry would have occurred in her body if she had not suffered the shock of her daughter's death. Whatever the answer to that may be, I doubt very much that she would have recovered so well and so rapidly if she had received no postventive therapy.

The parents had an extraordinarily good marriage. Many relatives gave them emotional support. The husband, more laconic and more stoic, was my principal co-therapist—although I did not forget his needs and saw him alone occasionally, when, for example, his wife was in the hospital.

Several months after the tragedy the parents seemed to be in rather good shape, physically and emotionally, everything considered.

They still had low-level grief, and no doubt always will. But what is most important for this discussion is that each of them has stated that the process of working through their grief was made easier for them and that the outcome was better and more quickly achieved (though not with undue haste) as a result of our postventive sessions, that something of positive value had been done for them, and that they felt that something of this nature ought to be done for more people who found themselves in similar situations.

Quite a few months have passed since the homicide, and the murderer has still not been apprehended. If an arrest is ever made, it will inevitably present renewed conflicts for the girl's parents, and a new question will be raised: Should someone also be concerned with the welfare of the parents of the accused after his arrest?

Most deaths occur in the hospitals, and the dying patient is often isolated and his awareness clouded by drugs. (We nowadays rarely see our loved ones die "naturally" at home—a common event a few generations ago.) The topic of death is an especially unpleasant one for medical personnel. Death is the enemy; death represents failure. It has been noted that physicians constitute one of the few groups (along with policemen and combat soldiers) licensed to take lives in our society; at the same time, however, relatively few physicians deal with dying patients (consider the vast numbers of orthopedists, dermatologists, obstetricians, pediatricians, psychiatrists, and other specialists who deal with conditions that only rarely result in death) and are therefore ill equipped to teach others about death comfortably and meaningfully.

From the point of view of the staff-patient relationship vis-à-vis death, there are essentially three kinds of hospital wards or services. They can be labeled benign, emergent, and dire. A benign ward is one on which deaths are not generally expected and the relationship between staff and patient may be of short or long duration: obstetrical services, orthopedic wards, psychiatric wards. A death on such a service is a sharp tragedy and cause for special self-examination. The reaction is both to loss of a person and to loss of control. The staff, typically, is not so much in mourning as in a state of shock at administrative and professional failure.

Emergency services are quite different. These are the hospital emergency room and the intensive care unit. Here death is not an uncommon experience, but the relationship between staff and patient is short-lived; the patient is hardly known as a person, and there is typically no time for meaningful interpersonal relationships to develop: the focus is on physiological functioning. Patients are often unconscious and almost always acutely ill. Death on such a service is not mourned as deep personal loss. It is a "happening," distressing but seldom totally unexpected. The staff—often self-selected—must be inured to the constant psychological toll of working in such a setting. The dangers are callousness, depression, and even acting out, especially in the forms of alcoholism, drug abuse, and heightened sexual activity.

It is the dire service that poses the most stressful problems for physicians and nurses. On such a ward the patients have grim prognoses (and are often doomed when they come to the ward, their illnesses diagnosed as cancer, leukemia, scleroderma, or whatever), and they remain there for an extended period of time, long enough for personal relations to be formed and for them to be known, loved, and then mourned as "real" human beings. Physicians practicing specialties dealing with fatal conditions—certain hematologists, oncologists, radiologists, and so on—know death all too intimately. Often there is not a week that is free of a death. The psychological stresses on such a service may be even greater for nurses, for they face the problem of giving intimate care, risking personal investment, and then dealing with loss.

Consider the following case: A thirty-year-old man was admitted to a hospital ward diagnosed as having leukemia in the terminal stage. He had an unusual combination of physical and personality characteristics that made him an especially "difficult" patient—not difficult in behavior, but difficult to see grow more ill and die: he was handsome, good-natured, alert; he had a keen sense of humor and was flirtatious with nurses, and through intermittent spells of depression and concern with dying he was remarkably brave, reassuring doctors and nurses and telling them not to worry about him or take it so hard. (It would have been much easier for them if he had been difficult in the usual sense—querulous, demanding, complaining; then they could have accommodated more easily to his death, with no great sense of loss.)

As a consultant on the ward, I visited him each day, having been asked to see him because of his depression. He talked openly about the topic of death and his fears of pain, aloneness, and loss, and specifically about his own death and its meaning for his wife and children. These sessions were, by his own account, very meaningful to him. It was the kind of death work that is not unusual in this kind of circumstance. But what was of special professional interest was the behavior, both before and after his death, of a nurse and of his physician.

The nurse, an exceptionally attractive young woman, grew to like the patient very much, with feelings that seemed to strike deeper than routine professional countertransference. I once witnessed a fascinating scene in which the nurse was massaging the patient's back while his wife stood stiffly off to one side; it was an interesting question as to who, at that moment, "owned" the dying patient. (But this nurse was extraordinarily good with the wife, taking her out to dinner, helping her with her young children, making, in a friendly and noncompetitive way, a number of practical arrangements for her.) After the young man's death there were several mourners for whom postvention was necessary, and the nurse was one of them. She grieved for one she had loved and lost, and her grief was sharpened and complicated by its somewhat secret and taboo nature.

The reaction of the physician in the case, a hematologist in his forties, was rather different but no less intense. We had planned to get together after the patient's death to discuss, I thought, interesting features of the case and possible plans for future collaborative efforts. Instead, he came into my office and announced that the young man's death had been the last straw. He was sick and tired of having all the doctors in the county dump their dying patients on him. He wondered how I could bear to see the young man every day while he was dying, and I countered candidly that I had wondered the same thing about him. Mainly he wanted a safe arena in which to vent his feelings about having had enough of death and dying for a while. Within a few weeks he followed his announced intention of making a major change in his professional life: he accepted a faculty position in a medical school in another state and in a specialty other than hematology.

In my own postventive work I have come to at least a few tentative conclusions:

1. Total care of a dying person needs to include contact and rapport with the survivors-to-be.

2. In working with survivor-victims of dire deaths, it is best to begin as soon as possible after the tragedy; within the first seventy-two hours if possible.

3. Remarkably little resistance is met from survivor-victims; most are willing to talk to a professional person, especially one who has no ax to grind and no pitch to make.

4. The role of negative emotions toward the deceased—irritation, anger, envy, guilt—needs to be explored, but not at the very beginning.

5. The professional plays the important role of reality tester. He is not so much the echo of conscience as the quiet voice of reason.

6. Medical evaluation of the survivors is crucial. One should be alert for possible decline in physical health and in overall mental well-being.

Three brief final points: Postvention can be viewed as prevention for the next decade or for the next generation; postvention can be practiced by nurses, lawyers, social workers, physicians, psychologists, and good neighbors and friends—thanatologists all; and a comprehensive mental health program in any enlightened community will include all three elements of care: prevention, intervention, and postvention.

4

THE POSTSELF

There is a way in which we can experience a form of death in life, a way that is seldom damaging to the psyche and is in most instances beneficial: we can examine our fears and hopes about our reputation and influence after death--about what we may call our "postselves." This sort of introspection is, of course, a popular pastime among children ("They'll be sorry after I'm dead!"), and at one time or another all of us have entertained fantasies, varying from the sentimental to the hardheaded, of witnessing our own funerals. Few of us utterly abandon thoughts of survival in some form or other. They are our fragile hopes of escape from total annihilation.

And since our death is an experience for others if not for us, almost everyone does "survive" his own cessation for *some* period of time in *some* minds—for perhaps two generations in about a score of minds. The topic of survival after death is enormous and interesting, covering a wide range of emotionally held positions and beliefs. Of the respondents to a recent death survey (see Appendix), 43 percent tended to believe in a life after death. I imagine it is obvious by now that I do not share their belief in any possibility of survival after death, except in accordance with my own special definition of "survival" as the postself. (The reader who wishes to pursue the notion of actual survival after death as it is conventionally understood can turn to Ducasse's *The Belief in a Life After Death* [1961]; Henderson and Oakes's *The Wisdom of the Serpent* [1963]; or the chapters on "Frontiers of Speculation" in Toynbee's *Man's Concern with Death* [1969].)

The concept of the postself is of an entirely different character from that of survival after death. It simply asserts that the individual who is going to die can, in the present moment, actively entertain notions (fantasies, thoughts, yearnings, dreams) about what the world will be like when he is absent from it. It is nothing more than what William Saroyan describes in *How It Is to Be:* "We talked of the dead as if they were not dead."

One can, in the present, savor his imaginings of what will transpire after he is gone, and these thoughts and images can influence important aspects of his behavior: the provision of life insurance for his potential survivors, his psychological investment in work that will not only survive him but continue to be identified with him, his preference for having children over remaining childless, for sons who will bear his name over daughters who will adopt the names of others—in short, his concern with his continued "existence" in the minds of others.

All this relates to fame, reputation, impact, "holding on." One asks oneself, "How long will my words and actions be remembered after my death? In what way? By how many people, and by whom in particular? Will they be influenced by them at all? If so, how?"

The postself needs to be distinguished from what Toynbee (1969) has called a "bonus life," what Maslow (1970) labeled a "post-mortem life," and what Keats, earlier, had called a "posthumous existence." These phrases were meant to convey the notion of added years of living after an episode that took one close to death, such as a severe coronary attack. Maslow (1970) wrote:

> After my heart attack my attitude toward life changed. The word I use for it now is the post-mortem life. I could just as easily have died, so my living constitutes a kind of extra, a bonus. One very important aspect of the post-mortem life is that everything gets doubly precious, gets piercingly important. You get stabbed by things, by flowers and by babies and by beautiful things—just the very act of living, of walking and breathing and eating and having friends and chatting. Every-

thing seems to look more beautiful rather than less, and one gets the much-intensified sense of miracles.

The period of survivial after an episode that could have been fatal is often referred to as "borrowed time." I have known half a dozen people who earlier had seriously attempted suicide (one cannot doubt that a man who shoots himself in the chest or through the roof of the mouth is serious) and fortuitously survived; most of them were greatly improved psychologically after this severest kind of trial by fire, at least for a while. They had done penance, sought death, and magically survived. But alas, they were unable to maintain that sense of wonder and special appreciation that Toynbee and Maslow glowingly described when they wrote of their own post-mortem lives—largely because they were bedeviled by the possibility of a return to the mental state in which they had tried to end their lives.

Unlike the post-mortem life, the postself relates to the concerns of living individuals with their own reputation, impact, influence after death—those personal aspects that still live when the person himself does not. These thoughts seem to give many people the comfortable illusion of escape from total annihilation. Maslow (1970) wrote: "Sometimes I get the feeling of my writing being a communication to my great-great-grandchildren, who, of course, are not yet born. It's a kind of expression of love for them, leaving them not money but in effect affectionate notes, bits of counsel, lessons I have learned that might help them." And Susanne Langer dedicated her book *Mind: An Essay in Human Feeling* "to them in whom I hope to live even to the great World Peace—my children and their children." Tom Farber simply entitled a recent book *Notes to My Unborn Son's Children.*

There is, of course, a broad range in the intensity of investment in the postself. At one extreme is the Nobel biologist James Watson's remark in the *New York Times Magazine* of August 18, 1968: "I believe in now—to hell with being discovered when you're dead." And Simone de Beauvoir, writing of her mother's death in *A Very Easy Death*, stated: "Religion could do no more for my mother than the hope of posthumous success could do for me. Whether you think of it as heavenly or as earthly, if you love life, immortality is no consolation for death." At the other extreme is the tragic hunger for immortality of Unamuno, the Spanish philosopher and writer (1954):

"For the sake of a name man is ready to sacrifice not only life but happiness—life as a matter of course. 'Let me die, but let my fame live! death is bitter, but fame is eternal.'" Unamuno was obsessed with immortality. "If we all die utterly," he wrote, "wherefore does everything exist?" And "If there is no immortality, of what use is God?"

It is interesting to note the parallel between Unamuno's statement and one of Freud's. Unamuno (1954), as far from the psychoanalytic orientation as a twentieth-century thinker can be, said, "It is impossible for us, in effect, to conceive of ourselves as not existing, and no effort is capable of enabling consciousness to realize absolute unconsciousness, its own annihilation." Freud's (1915) often quoted statement reads: "Our own death is indeed unimaginable, and whenever we make the attempt to imagine it we can perceive that we really survive as spectators. Hence the psychoanalytic school could venture on the assertion that at bottom no one believes in his own death, or to put the thing in another way, in the unconscious every one of us is convinced of his own immortality."

There are a number of ways in which an individual can continue to "live on" after his death:

1. In the memories of others. "We can make our actions count and endow our days on earth with a scope and meaning that the finality of death cannot defeat." (Lamont 1950)

2. By the active stimulation of others through one's works, e.g., art, music, books. "As with men, it always seemed to me that books have their peculiar destinies. They go towards the people who are waiting for them and reach them at the right moment. They are made of living material and continue to cast light through the darkness long after the death of their authors." (Serrano 1966)

3. In the bodies of others, through organ transplantation. Malcolm Muggeridge (1970) has a sardonic bit to say on this score:

> There is even a hope, turning into an assumption, that "Science" will somehow provide a remedy; from steadily expanding our expectation of life, find the means to prevent its termination ever. Keeping us on the road indefinitely, like vintage cars, by replacing our worn-out parts—new hearts, new lungs, new kidneys, new brains even, and above all, new genitals

as and when required. The everlasting life promised so dubiously by the Christians will then become an actuality. We shall, at last, be immortal. Or rather, some of us will. The supply of spare parts and the availability of skilled fitters is unlikely to stretch, initially at any rate, to more than a favored few, selected, presumably, among the richest, cleverest and most successful, with the *retardataires* furnishing the requisite organs.

4. In the genes of one's children. Hinton (1967) puts it this way:

> There is some biological immortality. It is more lasting than the short-lived remembrance of the deceased in the society to which he belonged. For a while personal memories and influences may linger and perhaps a lifetime's achievements will endure rather longer, but firmer assurance that personal life does not utterly perish depends on the individual's children. The elderly get comfort when they see their children and perhaps their children's children continuing in this life. They are links in a potentially immortal chain. Within the living bodies are the chromosomes, inherited from parents and ancestors and capable of infinite replication. Once the individual has safely passed on this genetic endowment he has contributed to his immortality. In accordance with much that goes on in nature, however, he himself need survive no longer. It is egocentric to hope otherwise.

Hinton would seem rather severe on spinsters, bachelors, and childless couples. But this feeling, the sense of what Erikson has called "generativity," is a very deep one, perhaps one of the most influential of human motives. In Volume 2 of his autobiography Bertrand Russell (1967) addresses himself to thoughts of this nature:

> Parental feeling, as I have experienced it, is very complex. There is, first and foremost, sheer animal affection, and delight in watching what is charming in the ways of the young. Next, there is the sense of inescapable responsibility, providing a purpose for daily activities which skepticism does not easily question. Then there is an egoistic element, which is very dan-

> gerous: the hope that one's children may succeed where one has failed, that they may carry on one's work when death or senility puts an end to one's efforts, and, in any case, that they will supply a biological escape from death, making one's own life part of the whole stream, and not a mere stagnant puddle without any overflow into the future.

5. And, philosophically, in the cosmos. Serrano, speaking of Hesse, says (1966):

> I sat on the grass across from the grave and thought of my friend, trying to recall his features, and to fix them in my mind as they had been before he had been carried down that immense river which vanishes completely in the sea and which disperses everything beyond the recall of memory. And then I remembered his words. "To die is to go into the Collective Unconscious, to lose oneself in order to be transformed into form, pure form."

In one's investment in his postself two orientations can generally be distinguished: self-orientation, the wish to be recognized for one's contributions, made up of such elements as narcissism, pride, vainglory, and egocentricity, not to mention "proper self-respect"; and service orientation, the wish to contribute to humanity, made up of selflessness, humility, compassion, and sacrifice. In most of us these two skeins intertwine; we have threads of both. We are givers and takers, selfish and selfless. R. D. Laing (1967) asks:

> Who is not engaged in trying to impress, to leave a mark, to engrave his image on others and on the world—graven images held more dear than life itself? We wish to die leaving our imprints burned into the hearts of others. What would life be if there were no one left to remember us, to think of us when we are absent, to keep us alive when we are dead? And when we are dead, suddenly or gradually, our presence, scattered in ten or ten thousand hearts, will fade and disappear. How many candles in how many hearts? Of such stuff is our hope and our despair.

Kaplan (1970) reflects on this issue in these measured terms:

> Parallel to the fallacy of the self's simple location in space is another with regard to its location in time. Just as the self is not contained in the skin, so also does its history not end with the dissolution of the skin. There is a perfectly reasonable sense of immortality which refers to a man's continuance in his children, his work, and his influence. When Whitman says of his *Leaves of Grass* "who touches this touches a man" he is not indulging in poetic license. The spirit that survives, in the present sense, is not a subtle kind of matter but a pattern or meaning which can be embodied and expressed in one material as well as in another. There *is* a dissolution of the self—"immortal" need not be taken to mean "everlasting"; the self ends as a wave dies down or as an inscription is eroded. The words of the Dhammapada have an unexpected relevance: "Those who are in earnest do not die, those who are thoughtless are as if dead already." Whether there is a continuance of the sense of identity is another question; who remains to say I? But it is still far from clear who it is that says I now.

Perhaps it is the poet, after all, who expresses this sentiment most pointedly. When W. H. Auden received the 1967 National Medal for Literature, he said in his acceptance speech:

> To believe in the value of art is to believe that it is possible to make an object, be it an epic or a two-line epigram, which will remain permanently on hand in the world. The probabilities of success are against him, but an artist must not attempt anything less. In the meantime, and whatever is going to happen, we must try to live as E. M. Forster recommends that we should: "The people I respect must behave as if they were immortal and as if society were eternal. Both assumptions are false. But both must be accepted as true if we are to go on working and eating and loving, and are able to keep open a few breathing holes for the human spirit."

I turn again to Herman Melville's convoluted life as an example and note that many of his more profound identity-seeking and death-focused labors—*Mardi, Moby Dick, Redburn, Pierre, Clarel, Billy Budd,* those works that sprang out of his deeper shaping and crushing concerns—expressed the concept that I have labeled the postself:

> [In *Redburn*]: Peace to Lord Nelson where he sleeps in his mouldering mast! but rather would I be urned in the trunk of some green tree, and even in death have the vital sap circulating round me, giving of my dead body to the living foliage that shaded my peaceful tomb.
>
> How much better would such stirring monuments be full of life and commotion, than hermit obelisks of Luxor, and idle towers of stone; which, useless to the world in themselves, vainly hope to eternalize a name, by having it carved, solitary and alone, in their granite. Such monuments are cenotaphs indeed; founded far away from the true body of the fame of the hero; who, if he be truly a hero, must still be linked with the living interests of his race; for the true fame is something free, easy, social, and companionable. They are but tombstones that commemorate his death, but celebrate not his life.

> [In *Moby Dick*]: It may seem strange that of all men sailors should be tinkering at their last wills and testaments. . . . After the ceremony was concluded upon the present occasion, I felt all the easier; a stone was rolled away from my heart. Besides, all the days I should now live would be as good as the days that Lazarus lived after his resurrection; a supplementary clean gain of so many months or weeks as the case might be. I survived myself.
>
> I now leave my cetological System standing thus unfinished, even as the great Cathedral of Cologne was left, with the cranes still standing upon the top of the uncompleted tower. For small erections may be finished by their first architects; grand ones, true ones, ever leave the copestone to posterity. . . .
>
> Immortality is but ubiquity in time.

And from his letters:

To Evert Duyckinck, April 5, 1849: "All ambitious authors should have ghosts capable of revisiting the world, to snuff up the steam of adulation, which begins to rise straightway as the Sexton throws his last shovelfull on him.—Down goes his body and up flies his name."

To Lemuel Shaw, April 23, 1849: "These attacks are matters of course, and are essential to the building up of any permanent reputation—if such should ever prove to be mine."

To Hawthorne, June 1, 1851: "What 'reputation' H. M. has is horrible. Think of it! To go down to posterity is bad enough, any way; but to go down as a 'man who lived among the cannibals'! When I speak of posterity, in reference to myself, I only mean the babies who will probably be born in the moment immediately ensuing upon my giving up the ghost. I shall go down to some of them, in all likelihood."

To Samuel Savage, December 10, 1862: "I once, like other spoonies, cherished a loose sort of notion that I did not care to live very long. But I will frankly own that I have now no serious, no insuperable objections to a respectable longevity. I don't like the idea of being left out night after night in a cold church-yard."

The concept of the postself is, of course, directly related to the notion of annihilation. Without these "breathing holes for the human spirit" we suffocate. And without the belief or the wistful hope that we shall survive—some part of us, in some way—our lives tend to lose whatever meaning we can manage to find in them. Not unexpectedly, Melville had a few things to say on these issues:

Israel Potter: "Few things remain. . . . He was repulsed in efforts after a pension by certain caprices of law. His scars provided his only medals. He dictated a little book, the records of his fortunes. But long ago it faded out of print—himself out of being—his name out of memory."

Redburn: "But all is now lost; I know not who he was; and this estimable author must need share the oblivious fate of all literary incognitos."

White-Jacket: "It is a good joke, for instance, and one often perpetrated on board ship, to stand talking to a man in a dark night-watch, and all the while be cutting the buttons from his coat. But once off, those buttons never grow on again. There is no spontaneous vegetation in buttons."

Pierre: "Unendurable grief of a man, when Death itself gives the stab, and then snatches all availments to solacement away. For in the grave is no help, no prayer thither may go, no forgiveness thence come; so that the penitent whose sad victim lies in the ground, for that useless penitent his doom is eternal, and though it be Christmas-day with all Christendom, with him it is Hell-day and an eaten liver forever."

From his letter to Hawthorne, June 1, 1851: "Though I wrote the Gospels in this Century, I should die in the gutter."

To cease as though one had never been, to exit life with no hope of living on in the memory of another, to be obliterated, to be expunged from history's record—that is a fate literally far worse than death.

2

VIEWS OF DEATH

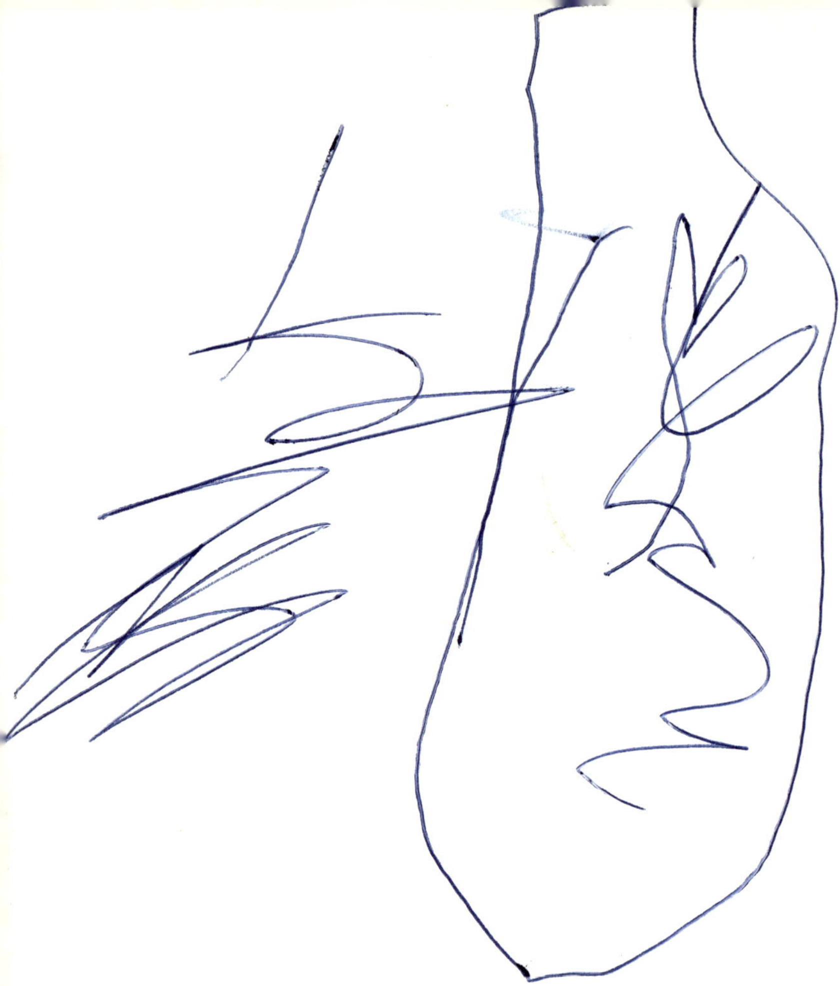

5

PERSONAL AND IMPERSONAL ASPECTS OF DEATH

The dual nature of death is its most characteristic feature. One can experience empathically the death of another but, paradoxically, cannot experience the death of himself. Thus all of our "experience" of death is indirect and, on that account alone, is the more puzzling and tantalizing.

Death is hardly a unique experience. Most people witness it more than once, and indeed for some people it can become almost a commonplace event; but no individual can himself experience death. If one could experience it, he would in fact not be dead. Listen to a renowned scientist-philosopher—a Nobel Prize winner who at the age of seventy-nine put a bullet through his own brilliant brain—Percy Bridgman (1938):

> There are certain kinks in our thinking which are of such universal occurrence as to constitute essential limitations. Thus the urge to think of my own death as some form of my experience is almost irresistible. However, it requires only to be said for me to admit that my own death cannot be a form of experience, for if I could still experience, then by definition it would not be death. Operationally my own death is a fundamentally different thing from the death of another in the same way that

> my own feelings mean something fundamentally different from the feelings of another. The death of another I can experience; there are certain methods of recognizing death and certain properties of death that affect my actions in the case of others. My own death is such a different thing that it might well have a different word, and perhaps eventually will. There is no operation by which I can decide whether I am dead; "I am always alive."

There are, then, two refractory images (or phenomenological aspects) of death: the private aspect, as one anticipates it for himself, and the public aspect, as one experiences the death of another.* The individual cannot ever experience himself as he is experienced by others; he can never see himself unconscious, hear himself snore, or experience his own death.

Kaspar Naegele—another suicide—in his extraordinary and tortured posthumous book, *Health and Healing* (1970), states: "Time binds into one familiar web the experiences of the body as a maturing and declining organism. In the West, we tend to think of life as a finite stream of events suspended between two radical occurrences, birth and death." It is interesting to contemplate that, from the individual's own point of view, neither death nor birth is an occurrence, radical or otherwise. We have already seen that death is not an occurrence inasmuch as no one is ever in a position where he cannot say, in Bridgman's words, "I am always alive." And so with birth there is no possibility of the statement "I am being born" or "I have just been born." Life is introspection. With the cessation of life there can be no introspection, hence no introspective death; and it is most unlikely that the infant can experience his own birth introspectively. Curiously then, neither birth nor death is an event—a living act, an undoubted deed—for the chief participant. It is an event only for others. The philosopher Ludwig Wittgenstein (1922) said it succinctly: "Death is not an event in life. Death is not lived through."

An interesting notion to contemplate is that if one cannot experi-

*I am attempting to make a distinction different from the one made by Elliot, who, in his recent remarkable book, *Twentieth Century Book of the Dead* (1972), distinguishes between "private death" and "those public deaths of battle slaughter, massacre, and slow, cruel siege [especially the] kind of public death . . . which arises from *macro-violence in the man-made environment.*"

ence his own death, it follows, logically, that he cannot experience his own dying. "Now, wait a minute," you might say, "I'll grant that I can't experience being dead, but obviously I'm still alive while I'm dying and unless I'm unconscious, I can experience that." The fact is that you can never be certain it is dying that you are experiencing. "Dying" takes its only legitimate operational meaning from the fact that it immediately precedes death. One may think that he is dying and then survive, in which case he will probably have experienced a number of things, but dying will not have been one of them. In my own life, I have twice believed that I was dying. Once I "drowned" in the Pacific—struggled and "died" and later regained consciousness on a rescue boat of whose life-saving presence I had been totally unaware. Another time, in a hospital, I was equally sure that I was dying, having been led to believe so by a physician friend who, inappropriately viewing me as the healthy and jovial colleague he had learned to expect rather than the frightened patient that I was, thought it a great joke to inform me that I was moribund. Both times I was wrong in my belief that I was dying. But the logical point remains: the one time when I shall have been right, there will be no "I" to experience the validating criterion. That is precisely what Bridgman meant when he said, "I am always alive."

One can, of course, keenly experience his *belief* that he is dying, and that experience can be deathly real. He can also experience the anticipation (with dread, curiosity, anxiety, or whatever) of what will happen after he is dead—what I call the "postself"—but these anticipations are, at the times they occur, always present-moment experiences of life.

All this is not to gainsay the fact that people are often correct in thinking that they are dying, because they do then die. During an extended period of dying (or supposed dying), a person, unless he is massively drugged or comatose, is very much alive. The behavior in dying is a psychologically consistent though often exaggerated extension of the individual's personality and life style. His idiosyncratic ways of coping, defending, adjusting, and interacting remain with him, coloring his inner life and characterizing his behavior. Feifel (1959) says: "A man's birth is an uncontrolled event in his life, but the manner of his departure from life bears a definite relation to his philosophy of life and death. We are mistaken to consider death as a purely biological event. The attitudes concerning it and its meaning for the individual can serve as an important organizing principle in determining how he conducts himself in life."

Thus ways in which an individual lives through his dying period reflect his personal philosophy, the quality of his personal adjustment, his sense of fruition and self-realization, and his notion of the kind of death that is "appropriate" for him. In a standard textbook on clinical medicine, Draper, Dupertuis, and Caughley (1944) succinctly state, "Each man dies in a notably personal way."

Death—some form of termination—is the universal ending of all living things; but only man, by virtue of his verbally reportable introspective life, can conceptualize his own cessation. And, before and after all, what is death? It is the absence of life. And what, then, is life? Life, *human* life, is the life of the self, the life of the mind. *Your* life is the full accounting of your personal diary—case history, memory bank, transient experiences of the present moment. Your life is the life of your mind as you look out on the world and reflect upon yourself. It is your internal dialogue and debate. As Murray (1940) says:

> A personality is a full congress of orators and pressure-groups, of children, demagogues, communists, isolationists, war-mongers, mugwumps, grafters, log-rollers, lobbyists, Caesars and Christs, Machiavellis and Judases, Tories and Promethean revolutionists. And a psychologist that does not know this in himself, whose mind is locked against the flux of images and feelings, should be encouraged to make friends, by being psychoanalyzed, with the various members of his household.

By "the life of the mind" as a definition for life itself, I mean not only those aspects of mind amenable to immediate conscious recall or use, but also the teeming world of the unconscious. On this subject, one cannot do better than to quote Murray again:

> The self-consciousness acquired in the brain is like a tiny coral isle growing in a sea of dreams (containing representatives of the body), which influence its every waking moment and sweep over it in sleep. To understand the mind, therefore, one must reach for much that lies beyond the range of consciousness. What a man does and says in public is but a fraction of him. There is what he does in private, and the reasons he

> gives for doing it. But even this is not enough. Beyond what he says is what he will not say but knows, and finally, what he does not know.

The idea of life as conscious process is best described by William James in his *Principles of Psychology* (1890), especially in the chapters "The Stream of Thought" and "The Consciousness of Self." It is the stream that counts, not the stream bed, for only in the current does consciousness flow; only the current bears movement and the manifestation of inner life. Better the smallest rivulet than the grandest arid canyon.

Death is the stopping of this life. To the question "How do you conceptualize your own death?" here is a typical response from a contemporary college student:

> I know I will die and be dead forever. I will have no memory, knowledge, enjoyment. Nor will there be any me who will not remember, not know or not enjoy. Nor will there be any void, vacuum or lacking in the place of memory, knowledge and enjoyment. Death is like the evaporation of a drop of water. The water itself will remain, become part of a cloud or part of an ocean and lose itself. There will be more drops, of course, but none of them will be me.

Bridgman said: "My own death is such a different thing that it well might have a different word." I propose that we use the word "cessation."

Cessation is the stopping of the possibility of any further conscious experience. It follows immediately upon what Melville called the last scene of the last act of man's play. In order to understand the word "cessation" as used here, a few additional terms will be defined in this context: termination, interruption, continuation, and altered continuation.

Termination is the stopping of the vital physiological functions of the body, including such relatively gross manifestations as heartbeat, and breathing. Physiological termination is always followed within a few minutes at most by psychological cessation. The converse, however, is not always true; that is, it is possible for (subjective) cessation to occur hours or even days before (objective) termination. For example, when a person's skull has been crushed in an accident, cessation occurs the instant he loses consciousness for the last time, but he might be kept alive in a hospital, and no one would dream of burying or cremating him so long as he was still breathing and there were brainwave patterns on the EEG. Had you been he, your life would have ended the second your brain was destroyed. Others—your family, hospital personnel, whoever—might see you alive in the hospital, but that could be an experience only for them, not for you. Because the moment of biological death is socially defined by termination, we need that concept even in a relatively psychological approach to death.

Interruption is the time gap during which there is a stopping of consciousness, but with the expectation, and usually the actuality, of further conscious experiences. It is, to use two contradictory terms, a kind of "temporary cessation." Dreamless sleep is perhaps the best example of an interruption state (Shneidman 1967, 1970). The opening sentence of Nathaniel Kleitman's *Sleep and Wakefulness* (1963) reads: "Sleep is commonly looked upon as a periodic temporary cessation, or interruption, of the waking state which is the prevalent mode of existence for the healthy human adult." Other interruptions include unconsciousness, stupor, coma, fainting, seizures, and anesthetic states from which the individual regains consciousness. Interruptions can last from seconds to weeks. The last interruption of a man's life is, by definition, cessation.

Continuation—William James's term—is the uninterrupted experiencing of the stream of temporarily contiguous events. Our lives are thus made up of a series of alternating states of continuation and interruption. As one would imagine, *altered continuation* implies the continuation of the stream of consciousness without interruption, but in a way different from the individual's usual style of functioning, for example, intoxication, drugged states, hypnotic states, malingering, role playing, spying, feigning, and even what might be termed "unplugging" (Shneidman 1970). Fascinating accounts of the unplugging function of play, the carnival, and the festival are given by Huizinga (1950) and Roger Caillois (1961).

6

ON THE DEROMANTICIZATION OF DEATH

One of the principal paradoxes of our society is that death is both a popular and a forbidden subject. The death of any well-known person is discussed at length; at the same time, the subject of death itself is largely avoided. A number of contemporary authorities have indicated that death has replaced sex as a taboo topic. Some have even called death pornographic.

Geoffrey Gorer in an essay called "The Pornography of Death" (1965) has written: "In the 20th century there seems to have been an unremarked shift in prudery; whereas copulation has become more and more mentionable, particularly in the Anglo-Saxon societies, death has become more and more unmentionable as a natural process."

In the foreword to Arnold Toynbee's remarkable recent volume *Man's Concern with Death* (1969), Robin Denniston says, "The idea of this book occurred some five years ago, at a time when the discussion of death really did seem like the new pornography—a secret and shameful matter totally opposed to the life-enhancing virtues of sex, love, freedom, and immortality."

In a most unusual confession, Malcolm Muggeridge (1970) writes, "Death has replaced sex as the forbidden subject." Later he adds: "For most contemporary minds the notion of death is hidden away, unmentioned if not unmentionable, as the Victorians hid away the notion of sex."

Herman Feifel (1963) says that "death is a taboo subject in the United States, surrounded by disapproval and shame." Richard Kalish (1965) states that "death is blasphemous and pornographic. We react to it and its symbols in the same way as we react to any pornography. We avoid it. We deny it exists. We avert our eyes from its presence. We protect little children from observing it and dodge their questions about it. We speak of it only in whispers. We consider it horrible, ugly and grotesque."

Why has death become taboo? Why is it considered impolite or unseemly to speak of it openly? Perhaps we can get some clues if we reflect on the characteristics of death. The outstanding fact about it is that it is inevitable and irreversible; but so are some aspects of life, so these characteristics do not especially distinguish it. I believe that the main reason it has become taboo is that it is essentially ineffable (i.e., incapable of being adequately expressed in words). For unlike any other aspect of life that we know, death is nonlife. What is special about death is its epistemological uniqueness. Any attempt to understand it stretches and troubles the mind, for our thinking cannot truly comprehend the concept of nonbeing.

"Death means the end of consciousness. Like dreamless sleep. Like before birth. No memories. Nothing." So says one contemporary youth, catching the essence of this aspect of death.

To contemplate death—that meeting-point of life and nonlife, that absence of a self, that absence of a future, that experience that is not experienceable—is it any wonder that this epistemological never-never land makes death seem forbidding and forbidden? Compared to death, sex is open and aboveboard. When prudery is removed and obscenity and sadism are excluded, when we can accept sexual intimacy as an enjoyable and natural part of the lives of all of us, then sex is no longer pornographic; it is a living human event like eating or walking. But it takes much more rigor of disciplined thought to contemplate and disentangle the experienceable and inexperienceable, the imaginable and unimaginable aspects contained in the thought "my death."

If death boggles the mind, resists final answers, defies logical analysis, withholds a satisfactory answer, then obviously it must be special, arcane, sacred—taboo. It forces us to recognize our essential aloneness and our ultimate powerlessness and insignificance. However crowded the ship—"Like a frigate, I am filled with a thousand souls" (Melville)—life is always a solo voyage in a vessel whose bottom is guaranteed to rust away eventually and may, in fact, fall out at any moment. Should we wonder, then, that we fear and hate the thought of it, and seek to *deny* the reality of it for us? And what is denied is akin to what is considered taboo. Wittgenstein (1922), discussing death, said: "Whereof one cannot speak, thereof one must be silent." There, perhaps, is the strongest interdiction on the discussion of death we are likely to bear in our time.

Romantic death is, of course, quite another matter. Think of the novelists who have invited readers to weep at the deathbeds of Camille and Beth and a host of other hapless heroines; of the composers who have beguiled opera audiences to respond to the melodious mortality of Mimi and Carmen; and of playgoers who feel cheated if a drama fails to give them a "good death scene." Much of the world seems "half in love with easeful death," at least with romantic death.

The romanticization of death has a long history in Western culture; it can be traced back at least a thousand years, and probably is much older than that. A classic example, Harlequin, stretches from the Italian *commedia dell'arte* of the 1500s to Agatha Christie's 1930 novel *The Mysterious Mr. Quin*—that's Mr. Harly Quin. Recently psychologists have described the Harlequin complex: death as the lover (McClelland 1963). However much we deny the reality of death, the tendency to romanticize it is still very much with us.

That the romanticization of death can be a powerful impetus is demonstrated in this account from a nineteen-year-old Harvard student (Hesse and Camus were the two authors credited by many current college youth with influencing their attitudes toward death more than any other authors):

> My suicide attempt of two years ago related closely to the writings of Herman Hesse. I cannot enter into any discussion of death without some personal reference and no incident seems more appropriate. My suicide attempt may now be regarded as analogous to Hans' experience, in Hesse's *Beneath the Wheel*, of the "exquisite torment" of love: ". . . the pain of love signified that the morning peace of his life had been broken and that his soul had left that land of childhood that can never be found again."

> I regard my suicide attempt as ultimately benevolent, despite the decided risk to life and self-estimation involved, the sapping of energy, the imposition on circumstances and plans. I have, as a result, a redefined conception of death; stated simply, I recognize that my fascination lay not so much with the prospect of death as with the dramatic consequences of dying. In the romantic tradition of courtly love, one may love his mistress, but, often enough, not his wife; insofar as removal imparts its own fascination, I can love my dying, but not my death.
>
> I now feel, as perhaps Hesse rediscovered in the course of *Steppenwolf* and *The Journey to the East,* that I cannot be my own "mirror" or my own "magic theatre," but must enjoy others and not prohibit them from enjoying me. Perhaps the attempt would have been avoided if the distinction between the romance of dying and the fact of death had been defined earlier.

Not only students, but professors too, write of the romanticization of death. The distinguished historian Arnold Toynbee (1969) turned to the topic of death as he neared eighty, and his description of the interpersonal aspects of death is unusually personal and moving. With passionate emphasis, Toynbee makes the point that death is essentially a dyadic (two-person) event, and that the survivor's burden is the heavier:

> The spectacle of insanity and senility has always appalled me more than witnessing or hearing of a physical death. But there are two sides to this situation; there is the victim's side, as well as the spectator's; and what is harrowing for the spectator may be alleviating for the victim. This two-sidedness of death is a fundamental feature of death—not only of the premature death of the spirit, but of death at any age and in any form. There are always two parties to a death; the person who dies and the survivors who are bereaved.
>
> If one truly loves a fellow human being, one ought to wish that as little as possible of the pain of his or her death shall be suffered by him or by her, and that as much of it as possible shall be borne by oneself. One ought to wish this, and one can, perhaps, succeed in willing it with one's mind. But can one

genuinely desire it in one's heart? Is it possible for love to raise human nature to this height of unselfishness? I cannot answer this question for anyone except myself, and, in my own case, before the time comes, I can only guess what my reaction is likely to be. I guess that if, one day, I am told by my doctor that I am going to die before my wife, I shall receive the news not only with equanimity but with relief. This relief, if I do feel it, will be involuntary. I shall be ashamed of myself for feeling it, and my relief will, no doubt, be tempered by concern and sorrow for my wife's future after I have been taken from her. All the same, I do guess that, if I am informed that I am going to die before her, a shameful sense of relief will be one element in my reaction.

This is, as I see it, the capital fact about the relation between living and dying. There are two parties to the suffering that death inflicts; and, in the apportionment of this suffering, the survivor takes the brunt.

In focusing on the importance of the dyadic relationship in death, Toynbee renders a great service to all who are concerned with death, particularly with suicide. The typical suicide is an intensely dyadic event. The crucial role of the "significant other" in suicide prevention has only in the past few years been adequately appreciated.

Although it is difficult to take a stance counter to Toynbee's, I believe that in emphasizing the dyadic aspect of death he seems to leap from a sentimental attitude of burden sharing in a love relationship—the noble husband's wish to save his beloved wife from the anguish of bereavement—to an unnecessarily romantic view of death itself. In cases of extremely sudden death, the total sum of pain is borne by the survivor (inasmuch as the victim cannot experience any of it). But in protracted dying the pain and anguish involved in the frightening anticipation of being dead may well be sharper for the dying person than the pain suffered then and afterward by the survivor. The algebra of death's suffering is a complicated equation.

For all his wisdom, I believe that Toynbee is indulging unduly in the romanticization of death. In my view, there is a very large need to *deromanticize* death, particularly suicide.

Certainly one of the most remarkable characteristics of man's psychological life is the enduring love affair that each of us has with his own consciousness. Trapped as he is within his own mind, man nurtures his conscious awareness, accepts it as the criterion for mediating reality, and entertains a faithful lifelong dialogue with it—even (or especially) when he "loses his mind" or "takes leave of his senses." Often man communicates with his mind as though it were a separate "other," whereas he is really communicating with himself. Indeed, the other to whom he talks is, in large part, what he defines himself to be. Death, like a cruel stepmother punishing noisy children, orders a stop to this fascinating conversation with the self. Death peremptorily decrees an abrupt, unwelcome, and final adjournment of what Henry Murray has aptly called "the congress of the mind." Should one traffic with one's greatest mortal enemy, rationalize its supposed noble qualities, and then romanticize it as a redeemer? I think not.

One can, as Eldridge Cleaver has said in his own naked way, put his "soul on ice" or, at the thermal extreme, one can bank the fires of his own passions (as Narcissus does in Hesse's *Narcissus and Goldmund*), but for a number of reasons it is next to impossible to defect from one's loyalty to his own consciousness. Being conscious is all one has. That is what one's life is. Consciousness defines the dimensions of life: its duration and its scope. The scope can be rich, or it can be a "partial death" that can come long before one's final cessation.

Many of our current attitudes toward death are unconsciously sentimental. The notions of "heroic death," of "generativity," of "wise death" in mature old age, are culture-laden rationalizations. It is pleasant to think that one can face the prospect of his own annihilation with true equanimity, but is it practical?

Although there are undoubtedly special circumstances in which some individuals either welcome their own cessation or are essentially indifferent to it, for almost everybody the heightened probability of death constitutes the most dire threat possible. By and large, the most distressing notion one can contemplate is his own cessation, however much religion and philosophy have attempted to soften this specter with notions of immortality. We must face the fact that death is the one act in which man is forced to engage. The word "forced" has a special meaning here. It implies that death, like torture, rape, capital punishment, kidnap, lobotomy, and other degradation ceremonies, is a form of coercion and impressment. The threat of being erased, of being reduced to nothingness, can be viewed reasonably only as the most perfidious of forced punishments.

In all this I do not believe that I am echoing Dylan Thomas's "Do not go gentle into that good night. Rage, rage against the dying of the light." I am saying rather that one should know that cessation is the curse to end all curses, and *then* one can, as he chooses, rage, fight, temporize, bargain, compromise, comply, acquiesce, surrender, welcome, or even embrace death. But one should be aware of the dictum: Know thine enemy. Death is not a tender retirement, a bright autumnal end of man's cycle, "as a chock of corn to his season." That notion, it seems to me, is of the same order of rationalization as romanticizing kidnapping, murder, impressment, the draft, or rape.

Nor does it mollify the arcane terror of death to discuss it in such honorific and beguiling terms as maturity, postnarcissistic love, ego integrity, or generativity, though one can only be grateful to Erik Erikson (1963) for the persuasive way in which he has made a generatic death sound ennobling and nearly worthwhile.

I wonder if it would not be better to understand generativity as reflecting pride in and gratitude to one's progenitors and pride and faith in one's progeny without the necessity of deriving any pleasure from one's own finiteness. There is (or ought to be) a reasonable difference between experiencing justifiable pride in what one has been and has created, and feeling an unwarranted complacency when one reflects that he will soon no longer be. Maturity and ego integrity relate to the former; the latter is supported largely by romantic rationalization that one's cessation is a blessing. Such a rationalization is no less than psychologically willing what is biologically obligatory. It would be more mature to bemoan this fact and regret it, as we bemoan and mourn most endings in our lives—divorce, the loss of a limb, and the deaths of others (Parkes 1970).

All this means that death is a topic for the tough and the bitter—people like Melville (in *The Lightning-Rod Man*): "Think of being a heap of charred offal, like a haltered horse burned in his stall; and all in one flash!" Or Camus, especially when he describes Meursault's strange burst of rage against God and death just before the end of *The Stranger*.

A look at another culture might throw some light on this problem. When I was in Japan a few years ago it seemed to me that one of the most pervasive religiocultural features of the country was the romantically tinged animism that infused religious thinking. The personal, spiritual feelings of the typical Japanese as he looked upon Mt. Fujiyama, Kegon Falls, Lake Chuzengi (the latter two are the sites of many suicides), or a cherry tree in its ephemerally beautiful bloom seemed totally different from the feelings of an average American viewing Mt. McKinley, Niagara Falls, Crater Lake, or a blossoming apple tree. The Japanese reverential closeness to nature, akin to deification, seems to lead to a special feeling toward death, a feeling I would have to call romanticization.

When I addressed a group of Japanese university students, one youth asked if I could give him any reason why he should not kill himself if he sincerely believed that by doing so he would become one with nature. The very question illustrates this animism and the romanticization of death. And I recall a young engineer who sat next to me on the Tokaido train and wrote out: "Cherry blossom is blooming quickly and scattering at once. Better to come to fruition and die like the blossom." He added: "We have had many great men among our forefathers whose deeds remind us of the noble characteristics of cherry blossoms."

Here, presaging Toynbee's statements by many centuries, is a translated passage from a twelfth-century Japanese short story called "An Account of My Hut": "There were other exceedingly unhappy occurrences. In the cases of husbands and wives who refused to separate during the famine, the ones whose affections were the stronger were certain to die first. This was because, whether man or woman, they thought of themselves second and gave to their beloved whatever food they occasionally managed to get. With parents and children, it inevitably happened that the parents died first."

In this country, the romanticization of certain types of homicide is an especially troublesome part of our national heritage. The honorifics go to the man with the gun. We have glamorized our rural bandits and our urban gangsters. The romanticized myth of the Western frontier, built around the image of the gunslinger, has set its homicidal stamp on our culture. The key problem for television may be less the effects of violence than the effects of the romanticization of violence.

This romanticization of death goes to the beginnings of our national history. One of our most popular symbols of the American Revolution is the Minuteman with his rifle—as though it were primarily guns that won the war. In 1818 John Adams (Bailyn 1967) wrote: "But what do we mean by the American Revolution? Do we mean the American War? The Revolution was effected before the war commenced. The Revolution was in the minds and hearts of the people; a change in their sentiments, of their duties and obligations. This radical change in the principles, opinions and affections of the people was the real American Revolution."

Perhaps more appropriate monuments at Lexington and Concord might have been statues of Paine and Jefferson, seated, with pens. Unquestionably these representatives would have implied quite different values and might have shaped our culture in a somewhat different fashion. Our recent inability to amend our gun laws in the wake of a series of catastrophic assassinations has been a national disaster and a grisly international joke, spotlighting our irrational tie to our essentially anti-intellectual legends of romanticized homicide. Our romanticization of killing in peacetime and of being killed in wartime (see, for example, the descriptions of the recipients of the Congressional Medal of Honor in Publication No. 1007 of the U.S. Government Printing Office, August 1943) is only one of the tragic paradoxes that result from our anachronistic values.

There is an obvious relationship between suicide and romantic notions of death. Individuals who are suicidal suffer, among their other burdens, from a temporary loss of the view of death as the enemy. This is the paradox and major logical fallacy of the suicide. He runs eagerly to his executioner.

Suicidal people mistake the foe for friend. They are traitors to reason. They attempt to rationalize death's supposedly lofty qualities and to romanticize death as the noblest part of love. Loyal people (that is, people loyal to life) are inured to nefarious propaganda leading to defection. One should not traffic with the enemy. Suicidal individuals have been brainwashed—by their own thoughts.

How could the deromanticization of death help suicidal people? Would it be beneficial to initiate programs aimed at deromanticizing death in our schools—courses in "death education"—or in our public media? In the treatment of the acutely suicidal person, what would be the effects of directing his mind to a view of death as an enemy? Would such a psychological regimen hasten the suicide, have no effect at all, or merely postpone the event? Youth, incurably romantic itself, requires, for its own life's sake, this deromanticization of death. In my own mind, the nagging question persists: Would not this type of effort, like practically every other earnest exhortation in this alienated age, itself be doomed to an untimely death?

But perhaps even more important is the question: How would the deromanticization of death reduce the number of criminal deaths—murder, massacre, genocide? Here I am a trifle more optimistic. If only we can recognize that three of our crushing national problems—the racial issue, the war in Indochina, and the threat of nuclear death—all contain the common element of dehumanizing others (and, concomitantly, of brutalizing and dehumanizing ourselves), then we might be able to reverse our present national death-oriented course.

3

NUANCES OF DEATH

7

SOMATIC AND TEMPORAL ASPECTS OF DEATH

At first thought, "death" is one of those patently self-evident terms, the definition of which need not detain a thoughtful mind for even a moment. Every mature person knows instinctively what he means by it. We don't need to consult the dictionary to know that death is the act of dying, the end of life.

Yet a moment's further reflection tells us that a lifetime might not be long enough to bring us to a full understanding of death. As we shall see, death is not the kind of concept that can be made concrete; rather it is one of those "horizonal phenomena"—Harold Garfinkel's felicitous term—that one can hope only to make more vivid or penetrable, appealing to commonsense reason and to the omnipresent emotional states that accompany our cognitive activities.

Take the question "What did Freud think about death?" One's best answer—as Litman (1967) has persuasively shown—might be the counter-question "What did he think about death *when*? In his teens? When he was thirty? Fifty?" Murray (1967) has superbly explicated these ages-in-relation-to-death for Herman Melville. In the same fashion, if one were to ask oneself "What is death?" one could not attempt an answer without specifying the kind of death he had in mind.

In recent years, especially in the past decade, several "kinds" of death have been proposed or defined. Even a cursory survey reveals more than a dozen kinds of death: anthropological, appropriate, brain, cardiac, civil, clinical, constructive, equivocal, partial, premature, presumptive, psychological, social, subintentioned. There are doubtless others. Some of these we have already discussed; others will be explored later. The various kinds of death with which we shall concern ourselves here are those that deal with the bodily functions and the times at which they cease: the somatic and temporal aspects of death.

SOMATIC DEATH

A. Keith Mant (1969) states that "somatic death is the cessation of all vital functions such as the heartbeat and respiration." He continues:

> Molecular or cellular death follows. Many cells in the body will continue to live for some time after somatic death. Muscles will respond, for instance, to electrical stimuli for up to two hours. The well-known death-stiffening of the body, rigor mortis, is due to cellular metabolism continuing after somatic death. Groups of cells may be removed from a body after death and kept alive, sometimes indefinitely, in tissue culture. The rate at which cellular death occurs varies in different organs. The more specialized the organ, the more rapidly its cellular death follows somatic death. The advent of organ transplantation surgery has greatly increased the importance of cellular death.

CARDIAC DEATH

"Cardiac death," to quote Robert Glaser (1970), occurs "when a patient's heart [has] stopped beating and he [has] ceased to breathe. Death might be verified by electrocardiogram." Glaser explains:

> Twenty-five years ago, when I was concluding my residency training, the decision as to when a patient was dead was a relatively uncomplicated one. When a patient's heart stopped beating and he ceased to breathe, he was pronounced dead. Death might be verified by electrocardiogram—the absence of electrical impulses on the tracing was considered decisive.

Under certain circumstances, the injection of such drugs as epinephrine (adrenalin) was carried out as a heroic measure but rarely produced more than a few additional spasmodic cardiac beats.

Since those days, however, medical science has developed new methods for support of both cardiovascular and respiratory functions. Thus, prompt cardiac resuscitation can restore a normal heartbeat in a significant percentage of patients stricken with a coronary occlusion and/or a lethal cardiac arrhythmia, such as ventricular fibrillation. Combined with mouth-to-mouth breathing or mechanical assistance from a respirator, cardiac resuscitation has in fact produced a marked reduction in the immediate mortality from acute myocardial infarction, and has improved the immediate prognosis of this common, life-threatening affliction.

Unfortunately, even short periods of circulatory failure, associated as they are with inadequate oxygenation of the blood, can result in irreparable brain damage. In such instances, despite the return of a normal heartbeat and subsequent adequate circulation of oxygenated blood, the patient no longer can function intellectually. In short, he is doomed to a vegetative existence. As measured by continued circulation and respiration, machinery can maintain life in such a person for long periods.

BRAIN DEATH

It is clear, then, that cardiac death is not necessarily irreversible. The heartbeat may be restored. But does this mean that the patient has avoided cessation? Glaser (1970) faces the issue squarely:

> In our inquiry into the meaning of death, we must differentiate between cardiac death and brain death. Because we are often slow to give up old concepts for new ones, this question may be debated for some time but, practically speaking, the answer is already in hand. Life and death, like day and night, are not absolute qualities, but move one into the other. Thus, death comes at different times to different tissues and organs, affecting cells of the brain in only a matter of minutes but less rapidly altering skeletal muscle and bone.

> Insofar as it involves organ donors and their rights, the technical question of death has been resolved, by popular consensus, in the recognition that the brain and not the heart is the seat of human life.

In 1968 Harvard Medical School's Ad Hoc Committee to Examine the Definition of Brain Death, chaired by Dr. Henry K. Beecher, reported its definition of irreversible coma, or brain-death syndrome (1968). In essence, the committee listed four criteria: unreceptivity and unresponsivity; no movements or breathing; no reflexes; and flat electroencephalogram ("At least ten minutes of recording are desirable, but twice that would be better"). The report explains:

> From ancient times down to the recent past it was clear that, when the respiration and heart stopped, the brain would die in a few minutes; so the obvious criterion of no heart beat as synonymous with death was sufficiently accurate. . . . This is no longer valid when modern resuscitative and supportive measures are used. These improved activities can now restore "life" as judged by the ancient standards of persistent respiration and continuing heart beat. This can be the case even when there is not the remotest possibility of an individual recovering consciousness following massive brain damage.

Glaser (1970) puts it most dramatically: "Although the beating heart remains for some a symbol of life and love, its role has been put into perspective scientifically. The brain is our master control; the heart is just a pump."

CLINICAL DEATH

Interestingly enough, the definition of clinical death comes to us from the behavioral scientists Sudnow (1967) and Knutson (1970). Knutson states that " 'clinical death' is the appearance of death signs upon physical examination." Sudnow introduces an important variable when he states that it is "important to note that the category 'DOA' is not totally homogeneous with respect to actual physiological condition. The same is generally true of all deaths, death involving, as it does, some decisional considerations, at least in its earlier stages." Clinical death is defined as a state evidencing signs of death, such as absence of audible heartbeat. In such a case, however, the patient is assumed to be potentially revivable and is treated as either "essentially dead" or "tentatively dead," depending on his assumed relative worth to society. The "tentatively dead" tend to be comparatively youthful and/or affluent, with access to extensive and personalized health care; the "essentially dead" are older and generally are dealt with differently. Sudnow (1967) states:

> Generally, the older the patient, the more likely is his tentative death taken to constitute pronounceable death. Before a twenty-year-old who arrives in the ER [emergency room] with a presumption of death, attached in the form of the ambulance driver's assessment, will be pronounced dead by a physician, very long listening to his heartbeat will occur, occasionally efforts at stimulation will be made, oxygen administered, and oftentimes stimulative medication given. Less time will elapse between initial detection of an inaudible heartbeat and nonpalpable pulse and the pronouncement of death if the person is forty years old, and still less if he is seventy. As well as can be detected, there appeared to be no obvious difference between men and women in this regard, nor between white and Negro patients. Very old patients who are considered to be dead, on the basis of the ambulance driver's assessment, were seen to be put in an empty room to "wait" several moments before a physician arrived. When a young person is brought in as a "possible," the ambulance driver tries to convey some more alarming sense to the arrival.

Sudnow is referring, of course, to patients whose admittance to the hospital has not been arranged by their own physicians. A "private" patient is generally accorded all possible measures to prolong his life, whatever his age.

BIOLOGICAL DEATH

References to biological death are made by several writers: Sudnow (1967), Brim, et al. (1970), Knutson (1970), and Kalish (1968). Knutson defines biological death as "the cessation of cellular activity." In general, the term is taken to refer to the physical symptoms associated with the culmination of physical death. Theoretically, this process occurs when measurable cell activity ceases; realistically, however, this is not an all-or-none process, since the cessation of the functioning of bodily parts occurs at differing rates, and these rates themselves vary from individual to individual. Thus the assessment of biological death is at the gross level. Kalish (1968) says:

> Physicians are likely to define two levels of physical death: biological death, which occurs when the organs cease to function, and clinical death, which occurs when the organism ceases to function as an organism. Biological death, however, is not an all-or-none proposition. . . . A recently reported case (*The New York Times*, September, 1966) described a girl whose heart, because of medical technology, continued to beat several days after her brain ceased to function; the time of her "death" was apparently debatable. Nor need the events be close in time. A part of the body may die, e.g. through an amputation, while the remainder may live for years. We also recognize that body cells die without causing clinical death; as a matter of fact, the human being has a changeover of cells every several years. Conversely, animal tissues have been preserved outside the living organism—far exceeding the life-span of the organism from which they were taken.

TEMPORAL ASPECTS

CONSTRUCTIVE DEATH

The term "constructive death" refers to the date on which an ill or injured person would have died if physicians had not intervened to keep him "alive" by technological means. Manning (1970), in discussing "liability risks to doctors and hospitals arising out of their administration or termination of care," notes that "the spurt in medical technology, with its potentials for extending life or terminating it," can be expected to increase these risks:

> Already mentioned in the context of the physician's liability is the impact on significant property distributions of the doctor's decision to terminate or not to terminate. It is possible to imagine legal contests between other litigants that might de-

pend upon a doctor's decision to continue or to terminate care. To put a simple case, suppose a husband and a wife are seriously injured in an automobile accident and the doctor decides to continue medical care in the case of one but terminate in the case of the other, with the result that one "dies" before the other. If under the terms of a will a devolution of property is determined by the sequence of the death, it is possible that in litigation involving interpretation of the will, one party might be arguing for recognition of the concept of "constructive death," or at least a constructive date of death—that date on which the injured person *would* have died had not the doctors intervened with massive medical technology. If that seems implausible, is it still implausible if the doctor removed vital organs from one spouse to maintain the life of the other? Is it implausible if care is terminated on one spouse, who dies at once, while intensive care is continued on the other, but it succeeds only in maintaining organic bodily functions and the patient never recovers consciousness or a capacity for communication; is it clear that for purposes of will interpretation one spouse should be considered to have died before the other?

PREMATURE DEATH

Premature death is generally thought of as death before one's time of productivity has ended, before one has either achieved his goals or abandoned them with dignity. Often people regard any death at a young age—roughly considered as any time between infancy and middle age—as premature. The heroic efforts of physicians to prolong life in youthful expirants reflects this notion, thus making "premature death" of a young person statistically less likely than that of a comparable older person. Riley (1970) has written:

> To the extent that the elimination of suffering and prematurity may permit a less threatening view of death, we may be on the verge of developing new attitudes and beliefs. Two aspects of possible change seem clear: one, individual; the other, social or interpersonal. For the individual . . . death is feared because it will cut short the achievement of goals upon which current self-esteem depends. In this view, death is not denied; rather, it is seen as a threat to the activity of life. The individual with

> such a view is expressing less a fear of death than he is of being "caught short." If death is not premature, it can be seen as an event characterized by personal fulfillment and individual dignity. In contrast, from the social point of view, the data emphasize not so much the threat that death poses for the individual and his identity, as the problems that death will create for the survivors. And here the evidence of concern is clear. In contemporary American society, concern with death prompts people to think of others far more often than it provokes anxiety about self.

Along these same lines Bailey (1970), discussing the economic and social costs of death, says:

> Several important points must be considered when we speak of death, particularly in relation to age and economic and social position. Age, alone, is especially significant; a person dying in his 20's, 30's or 40's has embodied within him a substantial amount of capital investment made by his family and society. Economic valuations of such persons tend, therefore, to be high and expenditures made on their behalf either by the individual or society to reduce the probability of premature death are likely to have a high payoff. In many respects, society acts quite rationally in providing health services for this age group through company-paid fringe benefits. Of course, society also exhibits just the contrary behavior in its willingness to freely expend the lives of its young men in wars.

These, then, are some of the kinds of death we must ponder and among which we must distinguish before we can ask ourselves "What is death?"

8

AMBIVALENCE AND SUBINTENTION

A passage from the unexampled paragraph from *Moby Dick* in the front matter of this book sets the tone for this chapter:

> And had you watched Ahab's face that night, you might have thought that in him also two different things were warring. While his one live leg made lively echoes along the deck, every stroke of his dead limb sounded like a coffin tap. On life and death this old man walked.

The simultaneous experiencing of contradictory psychological states—love and hate, dependency and autonomy, the drive toward life and the wish for death; in short, ambivalence—is perhaps the most fundamental concept of psychodynamic psychology.

Ambivalence is the simultaneous attraction toward and repulsion from a single object, person, or action. To understand ambivalence, one must first put aside his usual, ingrained Aristotelian patterns of thinking, which are essentially dichotomous. For Western man, something is either right or wrong, good or bad, pleasing or not pleasing. That is logical. But in his psychological life, one can often experience two seemingly contradictory states at the same time—pleasure and despair, for example. Psychological life, dealing as it does with emotions and other irrationalities, is not logical, nor is it meant to be.

Ambivalence is the most important psychodynamic concept to grasp if we are to understand our attitudes toward death. The concomitant movements toward the opposite goals of life and death are both genuine. One can swallow pills, genuinely wishing to die, and at the same time hope for rescue. The paradigm of suicide is one of the deepest ambivalence: to cut one's throat and to cry for help—in the same breath.

INTENTION AND UNINTENTION

When an individual plays a direct and conscious role in bringing about his own death, whether he experiences ambivalence or not, we may speak of that death as *intentioned.* Though the intention may be the same in all those who deliberately cause their own deaths, the motives behind the intention vary widely—as widely as the personalities and experiences of the individuals themselves. Let us consider some of these individuals:

First there is the *death-seeker.* A death-seeker is one who, during the time that label can properly be applied to him, wishes for an end to all conscious experience and acts in a way designed to achieve this end. The criterion for a death-seeker does not lie primarily in the method he uses—razor, barbiturates, carbon monoxide—but in the fact that *in his mind* the method will bring about cessation; and he commits the act in such a manner that rescue is realistically unlikely or impossible. During this period he is single-mindedly oriented toward death. Of course, an individual's orientations toward death shift and change over time. A person who was a death-seeker yesterday might today resist with all his strength any inducement to participate in activities that might cost him his life. It is known clinically that many individuals are suicidal for only a relatively brief period of time; if they can be given appropriate surcease and sanctuary, they will no longer seek death and will wish to continue to live.

Then there is the *death-initiator,* who believes that he will die in the fairly near future—a matter of days or weeks—or that he is failing, and, not wishing to accommodate himself to a new and less attractive image of himself, determines not to let it happen to him.

If death or decline is to occur, he wants to play the dominant role himself, at his own time and on his own terms. In investigations among a number of older hospitalized persons in the terminal stages of fatal illnesses, it was found that some, with remarkable and totally unexpected energy, succeeded in taking out their tubes and needles, climbing over the bed rails, lifting heavy windows, and throwing themselves to the ground several stories below. When the occupational histories of such individuals were studied, they were typically found to have had one thing in common: they had never been fired—they had always quit. Whether one quits or is fired, he ends up unemployed, but the role he has played in the process is quite different.

Next there is the *death-ignorer*. Consider the following suicide note: "Good-by kid. You couldn't help it. Tell that brother of yours, when he gets to where I'm going, I hope I'm a foreman down there; I might be able to do something for him." Although it is true that suicide notes that contain references to a hereafter are rather rare, it is also true that some people who kill themselves believe that one can bring about termination without actually effecting cessation. These people seem to ignore that fact that, so far as we know, termination always involves cessation. One can keep in mind that even those for whom no one ever "dies," but merely "passes on," will still say of the suicide that he "killed himself." Thus the death-ignorer—or perhaps better, the death-transcender—is a person who, from his point of view, effects only his termination and continues to exist in some other manner.

The concept of death-ignoring is necessary; otherwise we put ourselves in the untenable position of equating a man who shoots himself in the head in the belief that he will soon meet his dead wife in heaven with a man who takes a trip from one city to another with the purpose and expectation of being reunited there with his wife. Obviously, these two acts are so vastly different in their effects on others that they cannot even be compared. Therefore, regardless of any individual's conviction that killing oneself does not result in cessation but is simply a transition to another life, we must accept the notion that cessation is final as far as the human personality that we can know is concerned.

Last there is the *death-darer*, a person who, to use gamblers' terms, bets his life on a relatively low objective probability that he will survive. Regardless of the outcome, a person who plays Russian roulette—in which the chances of survival are only five out of six—is a death-darer. A person with very little skill as a pilot who attempts to fly an airplane or one with unpracticed coordination who attempts to walk along the ledge of a tall building may be classified as a death-darer. Thus, such a determination is based on not what one does, but the background (the skill, or lack of it, with which he performs the feat) against which he does it.

Two young people, the first a young man of twenty-one and the second a girl of nineteen, have written statements that have the ring of an authentic, if temporary, intention to bring about their own deaths:

> During my period of depression and drinking I thought to commit suicide by throwing myself on the third rail of the subway. Luckily, I passed out before I could even make it to the subway station. My reasons were that I wanted to end a long spell of depression. After this event, my desire for suicide ended and the depression began to leave.

> I did attempt suicide when I was fourteen. It was after my father had died; my mother had had a nervous breakdown and returned from the hospital she was in before she had completely recovered because we couldn't afford to keep her there any longer. She was a little crazy (very crazy, really) then. I wanted to go out one night and my mother forbade me. I saw the mess in my house, my mindless mother, and felt disgusted and sick. I wanted desperately to be free; I wanted to die and rejoin my father (I believed that through death we would be rejoined), so I jumped out of my bedroom window (about 25 feet up). I must have fallen correctly, because all I did was sprain my ankle. I ran (limped) off to a spot on a deserted hill to think and calm down. Then I got panicky that my mother must be worried, so I went back. She hadn't even known that I had left.

Most deaths, of course, are not suicides. They may come as a result of violence or of natural causes, but in either case, so far as the individual involved is concerned, they are unintentioned. An *unintentioned* death is any cessation, whatever its determined causes and its apparent conventional mode, in which the decedent plays no significant role in effecting his own demise. In these instances, death is due entirely to trauma from without (extrasomatic) or to biological failure from within (intrasomatic), neither of which is psychologically laden or induced. At the time of his cessation, the individual is "going about his own business" (even though he may be lying in a hospital bed) with no conscious intention of hastening cessation and with no strong conscious drive in this direction. Something from the "outside"—the outside of his mind—occurs. This "something" might be a cerebrovascular accident, a myocardial infarction, a neoplastic growth, some malfunction, some catabolism, some invasion by bullet or virus; whatever it is, for him it has lethal consequences. "It" happens to "him." Inasmuch as all that anyone can do in regard to cessation is to attempt to hasten or postpone it, one might suppose that the person who knowingly faces unintentioned death must inevitably wish to postpone it; but it appears that there are other possible attitudes—welcoming, accepting, resisting, disdaining, and more—all compatible with unintentioned death.

The *death-welcomer* is one who, although playing no discernible (conscious or unconscious) role in either facilitating his own cessation, can honestly report that he welcomes the end of his life. This is common among very old people, especially after a long, painful, debilitating illness.

The *death-accepter* may be distinguished from the death-welcomer by a nuance of passivity. The death-accepter is one who has accepted the imminence of his cessation and is "resigned to his fate." He may be relatively passive, philosophical, resigned, heroic, realistic, or mature, depending on the spirit in which this enormous acceptance is made.

Most of the time, most of us are *death-postponers.* A death-postponer is one who, to the extent that he is oriented toward or concerned with cessation at all, hopes that it will not occur in anything like the foreseeable future; the event must be staved off for as long as possible. (This death-postponing orientation should not be confused with the ubiquitous human fantasies of immortality; see Chapter 4.)

The *death-disdainer* is, in a sense, supercilious toward death. During those moments when he consciously contemplates cessation, he is disdainful of death and feels that he is above any involvement in the stopping of the vital processes that it implies. Most young children in our culture, despite their usual fears about death, are typically death-disdainers, as well they may be—for a while.

The *death-fearer* is one who is fearful of death and shrinks from anything related to death. He may be phobic about death. He fights the notion of cessation, seeing death as something to be feared and hated. This position may be related to wishes for omnipotence and investment in one's social and physical potency. Hypochondriacs, fearing illnesses and assault, are perhaps also death-fearers.

Imagine five people, all elderly men on the same ward of a hospital, all dying of cancer, none playing an active or unconscious role in his own cessation. Yet it is still possible to distinguish different orientations toward death among them: one wishes not to die and is exerting his "will to live" (death-postponer); another is resigned to his cessation (death-accepter); the third is disdainful of what is occurring to him and will not believe that death can "take him" (death-disdainer); still another, although taking no steps to hasten his end does at this point in his illness welcome it (death-welcomer); and the fifth is so fearful of death that he forbids anyone to speak of it in his presence (death-fearer).

It is, of course, possible to shout "Fire!" in the absence of a conflagration or "Stop thief!" in the absence of a crime. It is also possible, figuratively or literally, to shout "Suicide!" in the clear absence of any lethal intention. One expectation of shouting "Fire!" or "Stop thief!" is the mobilization of others. They are grab words; they give society (or certain members of society) no choice but to act in certain ways. An individual who cries "Suicide!" with a conscious absence of any lethal intention may be a *death-feigner*. A death-feigner is one who simulates what appears to be an advertent movement toward cessation: he drinks from a previously emptied iodine bottle or cuts his wrist superficially with a razor blade, with no lethal possibility or intent. Such a person may use the threat of suicide to manipulate other people—usually the "significant other" person with whom the individual is involved in a neurotic relationship.

SUBINTENTION

Subintention is a partly mysterious concept, resting as it does on the powerful idea of unconscious motivation. Unlike ambivalence, which characterizes the dual aspect of man's behavior, subintention exists in the unconscious depths of man's being. It is what Melville pointed to in Ahab's crew as "their unconscious understandings" and what he reified as "the subterranean miner that works in us all." Subintentioned acts, whether toward death or toward the expansion of life, are essentially movements toward goals that are not consciously recognized. They are life's maneuvers that well up out of unconscious motivations, and thus are subtle in their appearance and difficult to identify. Is smoking suicidal? Drinking? Driving? Skiing? These questions cannot be answered with a simple yes or no. The answer is "It depends," and it may depend on a number of factors, including the individual's orientations toward death and toward others in his life.

Subintentioned orientations toward death are, I believe, more common than most of us would care to recognize and characteristic of a large percentage, perhaps a majority, of all deaths. The subintentioned death is one in which the person plays some partial, covert, subliminal or unconscious role in hastening his own demise. The evidence for such a role might be found in a variety of behavior patterns: poor judgment, imprudence, excessive risk-taking, neglect of self, disregard of a life-extending medical regimen, abuse of alcohol, misuse of drugs—all ways in which an individual can advance the date of his death.

Traditionally, suicidal deaths are thought to be intentioned deaths while natural, accidental, and homicidal deaths are thought to be unintentioned. This distinction is not so clear in real life—or rather, real death. What happens in fact is that there are instances of all four types of death which can be subsumed under the category of subintentioned death, depending on the particular details of each case.

Many deaths certified as natural have a participative and subintentional quality about them. Many of us know of regrettable cases in which people with diabetes, peptic ulcers, cirrhosis, Berger's disease, or pneumonia have, through psychologically laden commission, or omission, precipitated their own deaths. In addition, "voodoo deaths," inexplicable deaths in hospitals (especially in surgery), and sudden declines in health can all be considered to be of this type. There is even a notion that the speed at which some cancers grow may be related to deep psychological conflicts.

And if some natural deaths are subintentioned (and thus not

entirely natural), many deaths certified as "accident" are even more so (and are thus not entirely accidental). The label "accident prone" would seem to be a contradiction in terms. Either an accident is an adventitious event or it is not an accident. If someone experiences a series of inimical events, they can hardly be thought to be "accidental." Sometimes, on the highway, we see a person drive as though he were afraid that he might be late for his own funeral; he may be hurling himself toward a subintentioned death. Many automobile fatalities are not quite accidents and not comfortably called suicides; they can be more meaningfully understood as subintentioned deaths.

Some suicides show aspects of subintention. (This is especially true of many cases of death certified as "probable suicide.") Indeed, the entire concept of subintentioned death is similar in many ways to Karl Menninger's concepts of chronic suicide, focal suicide, and organic suicide, except that Menninger's ideas have to do primarily with self-defeating ways of continuing to live, whereas subintentioned death is a way of stopping the process of living. Persons who die subintentioned deaths, may, in general, be said to have *permitted* suicide.

"Victim-precipitated homicide," a term introduced by Wolfgang (1959), conveys to us the concept of a fatal incident certified as homicide but which might better be described as subintentioned death. It is obvious that in some cases, especially among marriage partners, lovers, or close friends, the victim "sought to run more than halfway to meet his death"—like Rodney in *Moby Dick*. To provoke someone else to kill you is not to be an unequivocally unwilling victim; it is to participate indirectly, at some level of personality functioning, in the manipulation of one's date of death, and is thus a subintentioned event.

That individuals may play unconscious roles in their own failures and act in ways that are inimical to their own welfare are facts too well documented from psychoanalytic and general clinical practice to ignore. Often death is hastened by the individual's seeming carelessness, imprudence, foolhardiness, forgetfulness, lack of judgment, or some other psychological mechanism. Many patterns of mismanagement and brink-of-death living that result in death must be recognized as subintentioned.

Among those who play largely unconscious roles in inviting or hastening death is the *death-chancer*. The death-darer, death-chancer, and death-experimenter occupy positions on a continuum of expectation and possibility of death. If a death-darer plays a game that gives him only five chances out of six of continuing to live, then a death-chancer will require odds significantly greater than that, but his game will still involve a realistic risk of cessation. The games themselves—the methods used to court death—have little if anything to do with the motivations, conscious or unconscious, of the player. Most methods (razor blades, for example) can, depending on the exact place and depth of the cut or the calculated expectation of intervention by others, legitimately be thought of as either intentioned or subintentioned. Individuals who "leave it up to chance," who "gamble with death," who "half intend to do it," are the subintentioned death-chancers.

The individual who unconsciously brings about or exacerbates a physiological disequilibrium so that his cessation (which would, in ordinary terms, be called a natural death) is expedited may realistically be termed a *death-hastener*. He may earn this label either by the style in which he lives (the abuse of his body, usually through alcohol, drugs, exposure, or malnutrition) or, if he suffers from a specific physiological disorder, by the mismanagement or disregard of prescribed remedial procedures.

Closely allied to the death-hastener is the *death-facilitator*, who, while he is ill and his psychic energies are low, is somehow more than passively unresisting to cessation and "makes it easy" for death to occur. Some unexpected deaths in hospitals may be of this nature. The excellent work of Weisman and Hackett (1961, 1962, 1972) explores this area.

The *death-capitulator* is a person who, by virtue of some strong emotion, usually his great fear of death itself, plays a psychological role in effecting his termination. In a sense, he gives in to death, or he scares himself to death. This type of death includes voodoo death, and the type of death by suggestion reported by hospitals in the southwestern United States and other areas where patients whose experience with medical care has been limited or nonexistent believe that people who go to hospitals always die.

The *death-experimenter* is a person of quite a different sort. Often he lives "on the brink of death," not consciously wishing extinguishment, but—usually by excessive use of alcohol or drugs—seemingly wishing a chronically altered, usually befogged state of consciousness. Death-experimenters seek to remain conscious but unaware, to "see the world as the world's not." They will often experiment with increasing their self-prescribed dosages, taking some chances of extending the benumbed conscious state into a comatose state, and even taking some minimal but real risk (usually without much concern, in a kind of lackadaisical way) of extending the comatose state into cessation. When this type of death occurs, it is traditionally (but not necessarily correctly) thought of as accidental.

9

EQUIVOCAL DEATH

An equivocal death is one that cannot be neatly categorized as specifically natural, accidental, suicidal, or homicidal—that is, the *mode* of death is uncertain or unclear. This uncertainty usually arises between the modes of accident and suicide.

In the first chapter we read of the natural death (by lymphosarcoma) of a forty-eight-year-old woman and of the death, on the next day, of her only child, a twenty-eight-year-old son, in a fire in his apartment. A passing motorist saw flames coming from his building and called the fire department. The firemen found the doors to his apartment locked, had to break in, and found the apartment engulfed in flames and the young man dead on the floor near the door. Ruling out homicide and obviously eliminating a natural death, the question is whether his death was accidental or suicidal; the death was equivocal.

A sizable percentage of deaths—estimated between 10 and 15 percent—are equivocal as to mode, and most of these are eventually certified, without adequate psychological investigation, as accidental deaths. Obviously, some of these uninvestigated equivocal deaths are in fact suicide. The net effect is a gross underreporting of the number of suicidal deaths.

In the case of the young man who died in the apartment fire, an intensive investigation was conducted. The exact procedure and the results of this inquiry are reported in Chapter 12, on the psychological autopsy.

In general, when we exclude executions (legal and otherwise) and murders by hanging, deaths by hanging are usually considered to be instances of suicide. Hanging, like shooting oneself, ingesting a quick-acting poison, or jumping from a high place, is one of those methods by which one reaches the point of no return rather precipitously, and it is therefore one of the more assuredly lethal methods of suicide. Nevertheless, some cases of hanging are equivocal as to the mode of death. It is believed in psychiatric and forensic circles, for example, that when a man is found hanged, dressed in feminine clothes and/or with sexual materials on or near his body, his intention may not have been death, but rather a special orgasmic experience, heightened by the momentary cerebral anoxia that results from leaning into a rope. If he could come alive for a moment, he would be shocked to find that he was dead. Thus this type of death should accurately be counted as accidental rather than as suicide.

When equivocal deaths are investigated, the results are often enlightening from both a legal and a psychological point of view. Four such cases can be used as illustration, each involving a Caucasian male who died of asphyxiation associated with hanging. These victims came from different cities in the United States; in every instance the names and other details have been changed, but none of the essential features of the case has been altered.

Case 1 was a fourteen-year-old boy. Here is the investigator's report:

> The following is a report of the investigation of Gordon Kerr, a Caucasian male, aged 14. He had announced to his family that he was going to take a shower. The prolonged sound of running water prompted a 20-year-old brother to check the bathroom, and upon doing so he discovered the decedent hanging from the shower enclosure frame by a piece of heavy cloth belting. A city ambulance and a private physician were called. Subject was examined and pronounced dead.
>
> A preliminary police report suggested suicide although it was noted that there was no note or any circumstance outside the event to substantiate such a determination. The medical

examiner's office indicated "suicide/accident/undetermined."

Interviews with family members, friends, and decedent's football coach increased the probability of accident. Subject at 14 was a well-developed 160-pound youngster devoted to athletics, body building, and particularly football. He was a member of his school's squad and was scheduled to participate in a game the day following his death. Subject was known to engage in strenuous calisthenics designed to strengthen legs, shoulders, and neck muscles. These exercises included hanging momentarily with wide band cloths as "rope." Everyone associated with the subject is apparently seriously convinced that his death grew out of an exercise accident. In the absence of any indications of depressive or bizarre mood or any evidence of suicidal intent, it is difficult to reconcile a judgment of suicide.

On the basis of the information available and the apparent state of the subject's health and spirits, a recommendation of death by accident is suggested.

Case 2 involves a fifteen-year-old boy:

The deceased, Billy Bronze, was a 15-year-old Caucasian male. The parents were interviewed three months after the death. Both parents were affable, cooperative, and revealed little evidence of the depression that might have been expected after the death of their son.

The mother stated that a younger brother, age 11, and Billy had been playing in the garage on previous days, throwing a rope over a rafter and swinging by having the rope wound around their bodies and under their arms. The father said there had been much rope play in recent months, and on occasion the boys had put their feet in ropes and swung back and forth, standing on the rope.

Billy had had scarlet fever several months before but had completely recovered and was back in school, attending regularly except upon the day of his death, when his mother had told him that he would be going to the doctor's office for a regular checkup. The mother stated that Billy was somewhat disappointed that he would be missing school that day, but then agreed to take a bath preparatory to the scheduled physical examination. He had just finished bathing when she told him she was leaving to go next door for a few moments. She returned about ten minutes later. It was during this 10-minute interval that his death occurred.

The father states that the boy was found with his feet just a few inches off the ground and that he had stood up on a bar stool while attempting to put a rope through his arms in order to swing. He surmised that as Billy leaned forward to reach for the rope and put it around his arm, with the rope partially around him, he fell off the stool, and the rope, with a slip knot on it, caught him by the neck and strangulated him as he struggled to reach the floor. There were no notes; there was no evidence of depression or previous emotional difficulty.

The mother stated that the boy was singing in the bathtub, that she heard him through the door, that he appeared cheerful, and that he had plans for going to school in the afternoon and afterward delivering his papers as usual and then playing in the evening. Neither his sister, age 10, nor his 11-year-old brother revealed any material that would indicate that the boy was depressed.

The police notes stated that the boy had difficulty in school, but the father stated that he never was anything but a mediocre student and was a happy-go-lucky child who wasn't particularly concerned, despite the fact that his grades were poor.

The boy was in the habit of writing notes frequently to his mother, always telling her where he was going or what he was doing, and there were no notes found at the time of his death. The boy had taken out a $1,000 accidental death insurance policy with the local major newspaper (which he sold); the policy had been taken out at least several months prior to his death. Both the father and the mother stated they had been unaware that he had such a policy, but stated that they later learned that all of the boys who sold the newspapers took them out and that the money was given to the manager of the newspaper station directly by the boys. Billy was described as an easygoing child who made friends readily, did not worry, but was rather impulsive and thoughtless in his play, being little concerned about injury, and simply had not thought of the possibility that he might be accidentally hanged by this play. They deny that he had been self-destructive in any other way, nor was he a victim of frequent injuries, accidentally self-inflicted or otherwise. It is recommended that the mode of death be considered accidental, due to asphyxiation by hanging.

Case 3 is another instance of hanging in which, after investigation, a judgment of accidental death was recommended:

The deceased, George Palace, a 23-year-old bachelor, was found hanging from a tetherball pole on the playground of an elementary school. The toxicology report indicated ethanol (alcohol) absent. In connection with this investigation Palace's mother and brother were interviewed in our office on separate days. Three friends, his personal dentist, and his supervisor at work were interviewed by telephone. All of the individuals interviewed were responsible and the information is deemed to be reliable.

The deceased was described as an enthusiastic, high-spirited young man who loved life, knew how to live, and had the capacity to help others to enjoy life. He was a hobbyist, gadgeteer, and junior scientist. He had great interest in electronics, chemistry, good books, travel and transportation, pets, and body-building. He had extensive electronic equipment at home, as well as fish, a dog, a cat, and hamsters. He owned a motorcycle, a convertible, and a speedboat. He worked in electronics at a local corporation, was taking flying lessons, and was well on his way toward becoming a pilot. He was living life to the fullest and appeared to be enjoying every minute of it.

He was under no stress of an interpersonal or financial nature, was never depressed, and was never known to have spoken of suicide or death. On the contrary, things were going very well for him and he was to have received a substantial pay raise and to be assigned to a more interesting job at work. He was an independent person with many friends, although few close ones, and he was not interested in any female at the time of his death.

He had a bad back as a result of an auto accident in 1963 and he had been treated by a local physician. He was known to have exercised frequently, especially on the schoolgrounds, which were right next door to his home. On the morning on which he was found dead he was dressed for exercise and the position in which his body was found lends credence to his mother's strong belief that he was attempting to gain traction and relieve pressure on his back by slipping the tetherball rope around his neck and over his sweatshirt and pushing up and down with his legs while keeping his back to the pole. She believes that his foot must have slipped and that he lost consciousness and was thereby hanged accidentally.

On the last full day of his life he worked a full day at his job appeared his usual self, drove to the airport, and did a solo flight. He came home and went out to dinner with his mother and did not display any unusual emotion, but did say that he was feeling tired.

On the basis of the investigation of the psychological aspects of this case, a recommendation of accidental death is made.

Case 4 involves a twenty-nine-year-old attorney who was found dead in his combination den and gym, hanging from parallel bars. Perhaps the most noteworthy feature of the case lies in his wife's deep conviction that he was a victim of homicide, a belief she continued to hold for some time even after the death was officially certified as accidental. The medical examiner's report read as follows:

> This case, Robert Jameson, age 29, was reported by the police radio as an apparent suicide.
>
> According to the deceased's wife, Joan Jameson, she returned home at 8:00 A.M. today from her shift as laboratory technician at the University Hospital and discovered the deceased dead in the den of their home. Mrs. Jameson had left home at 11:00 P.M. last night, and the couple's child, a girl age six, was at home with the deceased. She stated that her husband was in a kneeling position on the floor with a small rope wrapped around his neck. This rope was threaded around the parallel bars and tied. The deceased had apparently wrapped the rope around his neck and got into a kneeling position, thus placing enough pressure on the rope to cause strangulation. There was a book beside him and a half-full bottle of whiskey.
>
> Mrs. Jameson stated that the deceased had had a drinking problem for several years, but after some counseling last year had controlled his drinking to a greater extent than previously.

> The deceased did not seem depressed last night, according to the wife, and no suicide notes were found.
>
> Mrs. Jameson cut the rope and placed her husband's body on the floor prior to calling the Police Department.

The medical examiner first labeled the death "undetermined," then "suicide." But that was not the final certification.

Subsequently the wife wrote to the medical examiner to say that her husband has been a hard worker, had made his own way through college and law school, and was not a quitter. She believed that people are not capable of doing things out of character, and on these grounds she completely rejected the idea of suicide.

A few weeks later the medical examiner's office replied to the decedent's wife: "We wish to advise you that the investigation into your husband's death has been completed. We have investigated and evaluated the circumstances and physical evidence surrounding the case. The cause of death of your husband has been determined to be accidental death by strangulation and not suicide."

A review of the case materials revealed that on the night of his death Jameson had been spending the evening in his den, drinking, watching television, and reading. The book he was reading was a pornographic novel with a good deal of sadistic content. The police estimated that the time of death was around midnight. The decedent was wearing no shoes or socks. There were no indications of any grasping at the throat or any scratches. They believed that he fell unconscious when the pressure of the rope cut off his oxygen supply, resulting in his death by asphyxiation. It was theorized that Jameson, while partially under the influence of alcohol, intentionally coiled the rope around his neck as a response to a sort of inspiration or suggestion he had received from the literature he had been reading. He became unconscious and died not intentionally, but as a result of this experiment.

When this theory was presented to Mrs. Jameson, she refused to believe it. She said that he was not capable of destroying anything, especially their life together. As for the book he had been reading, she said that he had wide-ranging reading habits; he preferred biography and fiction, but he would read anything. He was constantly buying or borrowing books. He was not a sexual pervert, he was simply an omnivorous reader. Further, she believed that he was murdered by someone who had come into the house that evening, and the act had been made to look like suicide.

In this case, the widow was faced with a most difficult and unusual dilemma. It can be stated this way: Imagine, a few years hence, her attempts to answer her child's questions about her father's death (keeping in mind that all her life she would continue to think about his death and try to make sense of it). Usually, survivors fight against a certification of the death of a loved one as suicide. In this case, the police and the investigators, after considering the facts, certified the death as accidental. If one is willing to rule out homicide—for which there was no evidence—then the widow was faced with the excruciating choice of explaining his death to the child either as an accident, with the inevitable implication that he had somewhat unusual erotic tastes, receiving sexual pleasure from leaning into a rope about his neck while indulging in sexual fantasies; or a suicide, with the implication that he must have been unhappy, despairing, or irrational. In effect, she was faced with choosing between two taboo-laden alternatives. In either case, the husband-wife relationship and father-daughter relationship would be tarnished or compromised. But the first alternative (accident) might be slightly more palatable to the surviving child. In time it would make her father understandable to her, and even though that understanding would have its unsavory aspects, it would have the advantage of avoiding the rejecting, irrational, and hostile implications of suicide, for which no totally satisfactory explanations could be offered.

When it was presented to her in this light, the wife seemed to understand the wider dimensions of the situation. She seemed to experience some relief and said, in effect, "Yes, I see what you mean. It might be better that way." It wasn't the answer she wanted, but since she could find no support for the notion of homicide, she seemed relieved to know that at least her late husband did not suffer a terrible despair that impelled him to suicide.

While uncertainty about the mode of death is most often associated with accidents and suicides, it is by no means confined to them. The mode of almost any death can be mistaken for another when the circumstances are sufficiently unclear, particularly when another person is involved in the death, thus raising the question of possible murder. Did the gun in the boy's hand discharge accidentally or did he shoot deliberately? If he shot deliberately did he mean to frighten or to kill? When the man parked his car on an incline, was he merely careless in setting the brake, or did he in fact deliberately release it just enough to permit the car to roll backward, crushing his wife? Such cases can be extremely difficult to resolve—difficult not only for the investigating authorities but frequently for the involved individuals themselves.

Such a case is nowhere more clearly illustrated than in a Japanese tale called "Han's Crime."* I shall retell it here, not only because of its relevance to the issue of equivocal death, but for the light it sheds on the differences in attitudes and legal procedures in other societies and other times, and on the necessity for compassion and insight if the truth is to be known and justice served.

Han was a young juggler who included in his repertoire a popular knife-throwing act. His wife, a very attractive young woman, would stand in front of a wooden board about the size of a door, and from a distance of approximately four yards Han would throw large knives so that they stuck in the board about two inches apart, following the contours of her body. As each knife left his hand, he would punctuate his performance with a staccato exclamation for dramatic effect.

One night, much to everyone's horror, one of the knives landed in his wife's throat instead of the board, severing her carotid artery. She died on the spot, and Han was immediately arrested.

Witnesses to the event included the director of the theater, Han's assistant, the master of ceremonies, some three hundred spectators, and a policeman who had been stationed behind the audience. Despite the large number of witnesses, no one could say whether the killing had been intentional or accidental.

*The story is the work of the Japanese writer Naoya Shiga (1956). While I have made a conscientious attempt to retain the flavor of the original, I have made a few changes in the wording of certain passages for the benefit of the modern American reader, who is unlikely to be familiar with certain social attitudes taken for granted in the Orient, and to whom the formal style of the essentially literal translation may seem so "literary" as to cast doubt on the credibility of the characters, and thus of the tale itself. It is difficult for us to believe, for example, that an uneducated man would use such words as "circumvent," "magnanimous," "retribution," and "stupefaction." Shiga himself, who died late in 1971 at the age of 88, frequently expressed distress at what he considered the inadequacies of the Japanese language, so we have reason to believe he would have been more pleased than not by minor alterations of this sort. In all essential respects, however, the tale remains unchanged, and I am grateful to Grove Press and to Ivan Morris who translated the story for permission to include it here.

The examining judge first questioned the director of the theater.

"Would you say that this was a very difficult act?"

"No, your honor, it's not as difficult as all that for an experienced performer. But to do it properly, you need steady nerves and complete concentration."

"I see. Then assuming that what happened was an accident, it was an extremely unlikely type of accident?"

"Yes, indeed, your honor. If accidents were not so very unlikely, I should never have allowed the act in my theater."

"Well, then, do you consider that this was done on purpose?"

"No, your honor, I do not, and for this reason: An act of this kind performed at a distance of twelve feet requires not only skill but at the same time a certain—well, intuitive sense. It is true that we all thought a mistake virtually out of the question, but after what has happened, I think we must admit that there was always the possibility of a mistake."

"It is your opinion, then, that this was simply a mistake rather than a deliberate act?"

"Well, I simply cannot say, your honor."

The judge was equally puzzled. Here was a clear case of homicide, but whether it was manslaughter or premeditated murder it was impossible to tell. If it was murder, it was a very clever one.

The judge next questioned the assistant, who had worked with Han for many years.

"What was Han's normal behavior?" he asked.

"He was always very correct, your honor. He didn't gamble or drink or run after women. Besides, last year he took up Christianity. He studied English and in his free time he always seemed to be reading collections of sermons—the Bible and that sort of thing."

"And what about his wife's behavior?"

"Also very correct, your honor. Strolling players aren't always the most moral people, as you know, and Mrs. Han was a pretty little woman, so quite a few men used to make propositions to her, but she never paid the slightest attention to that kind of thing."

"And what sort of temperaments did they have?"

"Always very kind and gentle, sir. They were extremely good to all their friends and acquaintances and never quarreled with anyone. But . . ." He broke off and reflected a moment before continuing. "Your honor, I'm afraid that if I tell you this, it may go badly for

Han. But to be quite truthful, these two people, who were so gentle and unselfish to others, were amazingly harsh to each other."

"Why was that?"

"I don't know, your honor."

"Was that the case ever since you first knew them?"

"No, your honor. About two years ago Mrs. Han became pregnant. The child was born prematurely and lived only about three days. That seemed to mark a change in their relations. They began having terrible rows over the most trivial things, and Han's face used to get white as a sheet. He always ended by suddenly turning silent. He never once raised his hand against her or anything like that—I suppose it would have gone against his principles. But when you looked at him, your honor, you could see the terrible anger in his eyes. It was quite frightening at times.

"One day I asked him why he didn't get a divorce, seeing that things were so bad between them. Well, he told me that he had no real grounds for divorce, even though his love for her had died. Of course, she knew how he felt and gradually stopped loving him too. He told me all this himself. I think the reason he began reading the Bible and all those sermons was to try to calm the violence in his heart and stop himself from hating his wife, because he had no real cause to hate her. Mrs. Han was really a pathetic woman. She had been married to Han nearly three years and had traveled all over the country with him, from one engagement to another. If she'd ever left Han and gone back home, I don't think she'd have found it easy to get married again. How many men would trust a woman who'd spent all that time traveling about, even if she'd been with her husband the whole time? A good woman is expected to marry a man who will provide her with a home, however humble, and that is her place—not traveling about and displaying herself before strange men. I suppose that's why she stayed with Han, even though they got on so badly."

"And what do you really think about this killing?"

"You mean, your honor, do I think it was an accident or done on purpose?"

"That's right."

"Well, sir, I've been thinking about it from every angle since the day it happened. And the more I think, the less I know what to make of it. I've talked about it with the master of ceremonies, and he says he can't understand what happened either."

"Very well. But tell me this: At the actual moment it happened, did it occur to you to wonder whether it was accidental or deliberate?"

"Yes, sir, I guess it did. I thought . . . I thought, 'He's gone and killed her.' "

"On purpose, you mean?"

"Yes, sir. But it was only a thought, on the spur of the moment. And the master of ceremonies says that *he* thought, 'His hand's slipped.' "

"Yes, but he didn't know about their everyday relations as you did."

"That may be, your honor. But afterward I wondered if it wasn't just because I did know about those relations that I thought, 'He's killed her.' "

"What were Han's own reactions at the moment?"

"He cried out, 'Ha!' just as he always did when he threw a knife. Like everyone else, I kept my eyes on the knife—to be sure it missed her, you know. Only this time it didn't. It landed in her throat, and suddenly blood was gushing out all over her. For a few seconds she kept standing there, then her knees seemed to fold under her and she swayed forward. The knife fell out as she collapsed on the floor, all crumpled in a heap. Of course there was nothing any of us could do—we just sat there petrified, staring at her. As for Han, I really can't describe his reactions, because I wasn't looking at him. It was only when the thought struck me, 'He's finally gone and killed her,' that I glanced at him. His face was dead white and his eyes were closed. The stage manager rushed to lower the curtain. When they picked up Mrs. Han she was already dead. Han dropped to his knees then, and he stayed like that for a long time. He seemed to be praying in silence."

"Did he appear very upset?"

"Yes, sir, he was quite upset."

"Very well. If I have anything further to ask you, I shall call for you again."

The judge dismissed the assistant and summoned Han himself to the stand. The juggler's intelligent face was drawn and pale; it was evident that he was in a state of nervous exhaustion.

"I have already questioned the director of the theater and your assistant," the judge said when Han had taken his place in the witness box. "I now propose to examine you."

Han bowed his head.

"Tell me," said the judge, "did you at any time love your wife?"

"From the day of our marriage until the child was born I loved her with all my heart."

"And why did the birth of the child change things?"

"Because I knew it wasn't mine."

"Did you know who the other man was?"

"I had a very good idea. I think it was my wife's cousin."

"Did you know him personally?"

"He was a close friend. It was he who first suggested that we get married. In fact, he urged me to marry her."

"I presume that his relations with her, if any, occurred prior to your marriage?"

"Yes, sir. The child was born eight months after we were married."

"According to your assistant, it was a premature birth."

"That is what I told everyone."

"The child died very soon after birth, did it not? What was the cause of death?"

"He slept with us in our bed, and he was smothered when my wife rolled partially over him, pinning his face against her breast."

"Did you believe she did that deliberately?"

"She said it was an accident."

The judge was silent and looked fixedly at Han's face. Han raised his head but kept his eyes lowered as he awaited the next question. The judge continued:

"Did your wife confess having had relations with her cousin or any other man?"

"Oh, no, and I never asked her about it. The child's death seemed like sufficient punishment for everything, and I decided that I should be as forgiving as possible, but . . ."

"But in the end you were unable to be forgiving?"

"That's right. I couldn't help thinking that the death of the child really didn't make up for anything. When I was apart from my wife, I was able to reason calmly, but as soon as I saw her, something happened inside me. When I saw her body, my temper would begin to rise."

"Testimony has been given that you thought about divorce. Is this true?"

"Yes, I often thought that I should like to have a divorce, but I never mentioned it to my wife. She used to say that if I left her she could no longer exist."

"She loved you, then?"

"No. No, she never loved me."

"Then why did she say she couldn't live without you?"

"I think she was referring to the practical business of living—having a place to live and enough to eat and that sort of thing. Her family's business had been ruined by her elder brother, and she knew that no serious man would want to marry a woman who had been the wife of a strolling player. Besides, her feet were too small for her to do any ordinary physical work, and of course she hadn't been trained to be anything but a wife."

"What were your physical relations?"

"I imagine about the same as with most couples."

"You have stated that your wife did not love you. But did she have any liking for you?"

"I don't think so, no. In fact, I think it must have been very painful for her to live with me as my wife. Still, she endured it. She endured it with a degree of patience almost unthinkable for a man. She used to observe me with a cold, hard look in her eyes as my life gradually went to pieces. She never showed a flicker of sympathy as she saw me struggling in agony to escape into a better, truer sort of existence."

"Why could you not take some decisive action—have it out with her, or even leave her if necessary?"

"Because my mind was full of all sorts of ideals."

"What ideals?"

"I wanted to behave toward my wife in such a way that there would be no wrong on my side. But in the end it didn't work."

"Did you ever think of killing your wife?"

Han did not answer, and the judge repeated his question. After a long pause, Han replied, "I often used to think it would be a good thing if she died."

"Well, in that case, if it had not been against the law, don't you think you might have killed her?"

"I wasn't thinking in terms of law, sir. That's not what stopped me. It was just that I was weak. At the same time I had this overpowering desire to enter into a truer sort of life."

"Nevertheless, you did think of killing your wife, didn't you—later on, I mean?"

"I never made up my mind to do it. But yes, it is correct to say that I did think about it once."

"How long was that before the event?"

"The night before—maybe it was the same morning. I couldn't sleep all that night."

"Had you been quarreling?"

"Yes, sir."

"What about?"

"About something so petty that it's hardly worth mentioning."

"Tell me about it anyway."

"It was a silly question of food. I get rather short-tempered when I haven't eaten for some time. Well, that evening my wife had been dawdling and our supper wasn't ready when it should have been, and I got very angry."

"Were you more violent than usual?"

"No, but afterward I still felt worked up, which was unusual. I suppose it was because I'd been worrying so much during those past weeks about making a better life for myself, and realizing there was nothing I could do about it. I went to bed but couldn't get to sleep. All sorts of disturbing thoughts went through my mind. I began to feel that whatever I did, I'd never be able to do the things I really wanted—that however hard I tried, I'd never be able to escape from the hateful life I was living. I was trapped, and it was my marriage that was keeping me trapped. I desperately wanted to find some way out, but there simply wasn't any. Because I knew that if I finally broke out—just walked out on her, you know—I couldn't live with the thought of what it would do to her, and I'd be just as trapped as I'd ever been. It would be worse, even—it would be like dying.

"And that's when the thought came to me: 'If only *she* would die! If only she would *die*!' What I really meant—and I knew it, your honor—was 'Why shouldn't I kill her?' The practical consequences of killing her meant nothing to me. I knew I'd probably go to prison, but life in prison couldn't be worse—could only be better—than the life I was living then. And yet somehow I had the feeling that killing my wife wouldn't really solve anything. It would only be a way of running away from my problems—like suicide. I must go through each day's suffering as it comes, I told myself; there's no way to get around that. That had become my true life: to suffer.

"By this time I had almost forgotten that the cause of my suffering was lying beside me. I was utterly exhausted, but still I couldn't sleep. I couldn't even think anymore. I just felt numb, and the idea of killing my wife gradually faded. Then I was overcome by the empty feeling of total defeat. All those fine resolutions to make a better life—and I didn't have the courage even to try! When dawn finally came I saw that my wife hadn't been sleeping either."

"When you got up, did you behave normally toward each other?"

"We didn't say a single word to each other."

"And you still gave no further thought to leaving her, when things had come to this?"

"No, no, your honor! I've already told you! I couldn't face what it would mean for her, and for me too. I was determined to behave toward my wife so that there would be no wrong on my side."

Han gazed earnestly at the judge, who nodded his head as a sign for him to continue.

"That day I was physically exhausted and of course my nerves were on edge. I couldn't sit still, and as soon as I'd got dressed I left the house and just wandered around. Constantly the thought kept returning that I must do something to change my life, but the idea of killing her no longer occurred to me. The truth is that there was a complete split, you might say, between my thoughts of killing her the night before and any decision to actually do it. In fact, I never even thought about that evening's performance. If I had, I certainly would have left out the knife-throwing act. There were several other acts I could have substituted.

"Well, the evening came and finally it was our turn to go on. I didn't have the slightest idea that anything out of the ordinary was going to happen. As usual, I demonstrated to the audience the sharpness of my knives by slicing pieces of paper and throwing some of the knives at the floor boards. My wife appeared on cue, bowed to the audience with her charming smile—at least she charmed the audience—and took up her position in front of the board. I picked up one of the knives and placed myself at the proper distance from her.

"That's when our eyes met for the first time since the previous evening, and all at once I realized the risk of doing this particular act that night. Obviously I would have to master my nerves, but I was so exhausted I simply couldn't manage it. It seemed that I could no longer trust my own arm. To calm myself I closed my eyes for a moment, and I could feel my whole body trembling.

"But the audience was waiting, and I had to get on with it: I aimed my first knife above her head. It struck about an inch higher than usual. My wife stretched out her arms, and I prepared to throw my next two knives underneath them. As the first one left the ends of my fingers, I felt as if something were holding it back; I no longer had the sense of being able to determine just where it would land. It was now really a matter of luck if the knife landed where it was supposed to; each of my movements had become self-conscious instead of spontaneous. It's like playing a stringed instrument. If you have to *think* about where to put your fingers on the strings—well, you have no business trying to give a concert.

"I threw one knife to the left of my wife's neck and was about to throw another to the right when I saw a strange expression in her eyes. She seemed to be frozen with fear—and suddenly so was I. I felt dizzy, as if I was going to faint. I forced my arm back and simply let fly."

The judge was silent, peering intently at Han.

"All at once the thought came to me, 'I've killed her,'" said Han abruptly.

"On purpose, you mean?"

"Yes. Suddenly I felt that I had done it on purpose."

"After that I understand you knelt down beside your wife's body and prayed in silence."

"Well, I knelt down. That part is true. It was a rather cunning device that occurred to me on the spur of the moment. I realized that everyone knew I was a Christian. But while I was pretending to pray, I was in fact carefully calculating what attitude to adopt."

"You are telling me that you in fact actually killed your wife deliberately?"

"No, but I thought then that I had. But I realized at once that I should be able to pretend it had been an accident."

"But you have already testified that you had lost your usual control—that it was now 'a matter of luck,' I believe were your words. Why then did you believe you had acted deliberately?"

"I had lost all sense of judgment."

"Did you think you'd succeeded in giving the impression it was an accident?"

"Yes, though when I thought about it afterward it made my flesh creep. I pretended as convincingly as I could to be grief-stricken, but if there'd been just one really sharp-witted person about, he'd have realized right away that I was only acting. Well, that evening I decided that there was no good reason why I shouldn't be acquitted; I told myself very calmly that there wasn't a shred of material evidence against me. It's true that a number of people knew how badly I got on with my wife, but if I insisted it was an accident, no one could prove it wasn't.

"And then a strange question came to my mind: Why did I myself believe that it had *not* been an accident? The previous night I had thought about killing her, but couldn't it be that very fact that made me think I'd actually done it deliberately? Gradually I came to the point where I didn't know myself what actually had happened and suddenly I felt so happy that I wanted to shout at the top of my lungs."

"Because you had come to consider it an accident?"

"No, that I can't say. Because I no longer had the slightest idea whether it had been an accident or not. But I felt it *could* have been. So I decided the best thing to do was to make a clean breast of everything. Rather than deceive myself and everyone else by saying it was an accident, why not be completely honest and say I simply don't know what happened? I can't swear it was an accident; on the other hand, I can't swear it was intentional. I can't plead either guilty or not guilty."

The judge too remained silent for a long moment before saying softly, reflectively, "I believe that what you have told me is true. I have just one or two more questions: Do you feel the slightest sorrow for your wife's death?"

"None at all! Even when I hated my wife most bitterly, I never could have imagined I'd feel such happiness as I do in talking about her death."

"Very well," said the judge. "You may stand down."

Han silently lowered his head and left the room. Feeling strangely moved, the judge reached for his pen. On the document that lay on the table before him he wrote the words indicating his verdict.

4

MEDICO-LEGAL ASPECTS OF DEATH

10

DEATH AND THE STATE

Until recent centuries official concern with death was largely confined to cases of murder. "Murder" has been defined in various ways by various societies throughout the world. What may be a clearcut case of homicide in one society at any particular time may be no murder at all in another place or at another time; but no society can tolerate murder as that society defines it. Unless chaos was to reign, murder called for official recognition and action. But deaths from other causes were of no concern to early governing bodies, and no official records of them were kept until the foundations of the modern nation-state were laid in the medieval period. The political state had reasons for keeping track of its citizens that smaller, more loosely organized societies did not: a state requires large revenues to finance its expanded administrative operations and wars, and it is the citizens that provided the revenues. The state's authorities thus began to take a keen interest in births and deaths within its jurisdiction—its "vital statistics." (The very word "statistics" has its origin in the Latin *statisticus,* "of politics," which in turn was derived from *status,* "state.") Yet even then the initial interest seems to have been in cases of murder or possible murder, including self-murder.*

*A first-rate tabular overview from ancient Egypt to 1967, the "Chronology of Important Events in the Development of Civil Registration and Vital Statistics," appears in Logan (1969).

One of the first reports of a document resembling a death record refers to the "rolls" that were collected and kept by the medieval English coroner. Though the coroner's office is generally assumed to have been created in 1194 (Gross 1896, Hunnisett 1961), there is some evidence that the position dates back to Norman times (Waldo 1910). The duties of the coroner are set forth in the Magna Charta of 1215 (Gross 1896). The coroner was the guardian of the crown's pleas—that is the origin of the word, *custos placitorum coronae*—and thus was involved with keeping written records. When he was summoned to court or ordered to act upon a writ, the coroner brought his "rolls." He was particularly concerned with sudden death, because if it were a case of murder or suicide, the property of the murderer or the suicide was forfeited and went to the crown. Thus there was intense interest in the way in which a death was certified.

These rolls may be regarded as crude death documents. They included the name of the deceased, the date and location of his demise, and a short explanation of the cause of death. The account of the cause was limited to a description of murder or accident. No medical explanation was offered. The only reference to age seems to have been a distinction between child and adult. Often some reference was made to family lineage. These documents were regarded as official. Indeed, any divergences from the "truth" resulted in penalties even for the coroner.

Here are a few representative rolls from the thirteenth and fourteenth centuries (Gross 1896):

> Bedfordshire County, April 8, 1268
>
> It happened in the vill of Renhold on Easter Day in the fifty-second year [of the reign of Henry III] about the hour of terce that a child fell into a well while his parcnts were at church and was drowned. His sister found him and produced two sureties.
>
> Inquest was made before Simon Read, the coroner, by four neighboring townships, to wit, Renhold, Wilden, Barford, and Goldington; they say that they know nothing except as is aforesaid.
>
> Gloucestershire County, February 25, 1298
>
> It happened at Stonehouse on Monday next before the feast of St. Martin the Bishop in the twentieth year of the present king that John Gabb of Stonehouse was found dead in a rivulet

called Littlere in the lordship of Stonehouse. The first finder was a poor stranger called Robert Lockyer; the pledges [for his appearance] before the itinerant justices, etc. are Walter Summerfowl and Walter Cowherd. The body was viewed by the aforesaid coroner on the following Wednesday.

Inquest was taken [at Stonehouse] before the said coroner on the oath of John Baker, Robert Walsh, John Stonehouse, and Walter Walsh, jurors of the township of Stonehouse, and Robert Daniel, Walter Prior, Thomas Moorcock, and Henry Maldon, jurors of the township of Stanley and Bernard French, Robert Webb, Robert Townsend, and William Amfrey, jurors of the township of Eastington. They say on their oath that the said John stood leaning against a willow tree, and while sleeping in this position, he fell into the said rivulet, and thus died by misadventure; and [that] they can ascertain nothing more. And he had in goods certain farm stock worth forty pence, which goods remain in charge of the township of Stonehouse.

Oxfordshire County, November 6, 1396

It happened on the morning of Tuesday next after the feast of St. Matthias the Apostle in the twenty-sixth year of King Edward that Thomas Churchey of Iffley died in Richard Pickard's house in the parish of St. Peter-in-the East, and he was viewed forthwith by the coroner aforesaid, and he had two mortal wounds almost through the body. Inquest was taken on that same day before the said coroner by three neighbouring parishes, to wit, St. Mary, St. Mildred, and All Saints. And all the jurors of that inquest say on their oath that on the preceding Sunday the said Thomas came into the High Street opposite the church of St. Mary the Virgin on his way to Iffley, and Roger of Brecham clerk came there, and struck the said Thomas with a long knife almost to the heart; and Henry of Brecham clerk struck Thomas with a sword almost through the body. Of these wounds the said Thomas died on the Tuesday aforesaid, but he had all the rites of the church. And Roger and Henry fled, so that they could not be attached owing to the large number of other armed clerks then and there present, nor could any of their chattels be found.

Gloucestershire County, November 26, 1398

It happened at Maisemore on Monday November the twenty-

> first in the twentieth year of the present king that Robert Hendy of Maisemore was found dead in a place called Abbotspool in the lordship of Maisemore. The first finder was Thomas Blake of Maisemore; the pledges for his appearance before the itinerant justices, etc. are Richard Hendy and Thomas West. The body was viewed by the aforesaid coroner on the following Wednesday.
>
> Inquest was taken [at Maisemore] on that Wednesday before the said coroner on the oath of Richard Fulmer, John Fisher, William Carter, John Style, William Piper, John Hull, John Taylor, William Code, Walter Gifford, Walter Kemp, John Stowell, and Richard Roberts, jurors of the townships of Maisemore, Over, and Highnam. They say on their oath that the said Robert knelt down to drink of the said water because he was ill and suffering from fever, but his hands failed to maintain their hold, and, falling into the water, he was drowned by misadventure. [The jurors] can ascertain nothing more. He has no goods. Further details are in the schedule sewn to this record.

The coroners' rolls served several purposes, financial as well as judicial. They were used in court proceedings when a criminal prosecution or the allocation of property was involved. In cases of appeal, the rolls were used as testimony to affirm or deny allegations. These records contained statements that often led to severe punishments. Punishable crimes included the burial of a body before it had been viewed by the coroner, and undue delay in summoning the coroner. Further, when the rolls stated that an accused murderer had fled, his property was confiscated and his heirs had no redress. It sometimes happened that a review of these records led to unexpected punishments. Occasionally punishment fell upon the coroners themselves, when irregularities in the execution of their duties were disclosed. Sheriffs might suffer penalties if the records revealed that the borough courts had illegally exceeded the bounds of their jurisdiction. Although the coroners' rolls were "of record," a check often revealed deficiencies, such as the omission of a case that a coroner was sworn to record or of specific details of a case presented by the jurors of the court.

The coroners' rolls remained the nearest approximation of the contemporary death certificate until around 1527, when "bills of mortality" began to be collected and published in London. Publication of these bills was somewhat haphazard at first, and they made no mention of cause of death unless the death was due to bubonic plague. After the bills began to be published on a regular basis in 1592, they gradually came to list the causes of all deaths. Weekly issuance began in 1603, following a mass epidemic of plague (Kargon 1963).

Unfortunately, the data contained in these bills were not very reliable because of the method of collection. Data were generally gathered by old women who took on the job as a means of earning the price of a drink or two, and any medical knowledge they had consisted of the home remedies common at the time. They were paid to obtain information about deaths and to quarantine the sick. The parish clerks received the findings on Tuesdays and were able to issue copies to the public by 10 A.M. on Thursday. Commoners perused these bills in hopes of finding news of some unusual occurrence that could be used as a subject for gossip; the well-to-do were concerned with increases in sickness and death during plague times, so they might judge whether or not it would be wise to leave the city for a while; and tradesmen observed these sheets with an eye to business prospects.

In Elizabethan England parish records of death yield a vivid picture of the vicissitudes of daily life (Forbes 1970). Forbes, writing of "life and death in Shakespeare's London," cites interesting descriptions of deaths from the record books maintained by the clerks of the parish of St. Botolph Without Aldgate from 1558 until 1625 and beyond. (These records are now in the Guildhall Library in London.) Deaths are cited as having been due to plague, "great age" ("Joane Blackborne a poore widow, reported to be above an hundred yeares old [November 2, 1622]"), and accident ("John Midleton, in Ship Alley, who by a fall from a scaffold caught his death [April 8, 1612]"). Many accidental deaths were by drowning: "[A boy] was drowned in Goodmans fields in a Pond, playing with other Boyes there and swymming [May 19, 1615]." Violent deaths also included infanticide, homicide ("This Ellen Hunte and Alice hir Mayd were both Cruelly murthered By one Morgan Colman who lodged in hir house, 27 Aprill [1609]"), executions (the bodies were properly buried), and suicide (the bodies were put in the ground without ceremony, sometimes at night at a crossroad after a stake had been driven through the heart: "Agnis Miller wieff of Jacob Miller who killed selfe with kniff was putt in the ground [August 27, 1573]"). In these fascinating vignettes one can see the beginnings of a classification of the modes of death: disease and plague, accident and drowning, violent deaths by homicide, execution, and suicide. A taxonomy of death is obviously implicit in these reports.

In 1662 John Graunt, a London tradesman, published a small book of "observations" on the bills of mortality that was to have great social and medical significance. By this time the weekly bills were consolidated at the end of each year, and a general bill for the year was published. Graunt separated the various bits of information contained in these annual bills into categories—at first births and deaths subdivided by parish and later also by sex—and organized them into tables. When the available data on deaths were believed accurate, Graunt then focused on individual causes of death. He next turned to the subject of population estimation. Finally he constructed a mortality table, the first attempt to organize data in this manner. Of greatest significance was his success in demonstrating the regularities that can be found in medical and social phenomena when one is dealing with large numbers. Thus John Graunt demonstrated how the bills of mortality could be used to the advantage of both the physician and government (Kargon 1963).

In 1741, the science of statistics, as it is known today, came into existence with the work of a Prussian clergyman, Johann Süssmilch, who made a systematic attempt to correlate "political arithmetic," or what we now call "vital statistics." From this study came what was subsequently termed the "laws of large numbers," which permitted extended use of the bills of mortality to supply important data in Europe as well as in the American colonies. Cassedy (1969) says that Süssmilch's "exhaustive analysis of vital data from church registers . . . became the ultimate scientific demonstration of the regularity of God's demographic laws."

In the American colonies, for many years no provisions existed anywhere for anything comparable to the London bills of mortality (Cassedy 1961). Though at first there was no necessity for detailed records, since communities were small, something better than hearsay or fading memories was needed as towns grew larger. The birth of newspapers in the British sections of America, about the year 1700, provided a means of remedying this situation. Some editors went to the trouble of obtaining information from local church and town records. They accumulated long lists to which summaries were added, which together were loosely called "bills of mortality." Issued in a variety of publications, the bills became the earliest systematized American death certificates.

Meanwhile, back in England, "the London Bills of Mortality remained among the eternal verities for Englishmen" (Cassedy 1961). Londoners were sure to find the bills on sale in each parish each week. Bills with lists of cases of various diseases printed on the back brought twice the price of the regular bills. The American colonists had relatives or friends send copies to them: "Just as John Graunt had found the English doing, the early Americans used the bills as grist for conversation if for nothing else" (Cassedy 1961). Few in the colonies even knew of Graunt's statistical applications to the bills of mortality for years after his book was published. Not until William Douglass commented upon statistical method in his history of the British colonies in North America, published in 1751, did the colonists begin to make use of Graunt's methods.

The fact that no colonial bills were published during the seventeenth century may be attributed to printing priorities and especially to lack of legislative requirements. Though a printing press had begun operation in Cambridge, Massachusetts, in 1638, the next half century saw the introduction of only four more. These few presses were generally kept running at capacity turning out government documents. Though the government of each colony required the registration of all vital statistics, none of them made any provision for the information to be published. Thus the publication of anything resembling the London bills was left to private enterprise. Despite the spread of colonial newspapers and the existence of a few church bills of mortality, British publications remained the colonists' best sources of vital statistical information (Cassedy 1961).

The use of such nonstandardized death records continued into the nineteenth century. Recognition of the need for informed medicolegal investigation in England led to a series of reforms aimed at improving the quality of death registration. In 1836 Parliament enacted a bill requiring the recording of all deaths. Under the terms of this act, notification of the coroner was not required unless the cause of death was one included in a special category. This system of death certification and registration was to ensure that those special deaths that came under the coroner's jurisdiction were in fact reported to him. The new certification system was designed to utilize the data on causes of death for statistical purposes as well as to prevent criminal practices. The medical explanation of the death was the essential information required for statistical determination. A curious aspect of the law, however, was that doctors were specifically required to include no information on the mode of death on the death certificate.

The English 1836 registration act was amended in 1874 to require that personal information on the death be submitted to the registrar of the district within five days of its occurrence by the nearest relative of the deceased who was present at the death or in attendance during the last illness. A fine was to be levied for noncompliance. The new law also required a registered medical practitioner present during the last illness to complete a certificate stating the cause of death to the best of his knowledge and belief.

In 1893 a "Select Committee on Death Certification" of the House of Commons attempted to correct the shortcomings of the previous legislation, particularly "the carelessness and ignorance of the persons certifying, the absence of medical attendants during the last illness and the indefinite character of the disease itself" (Abbott 1901). That committee made a series of ten recommendations, the most important of which were the following four:

> That in no case should a death be registered without production of a certificate of the cause of death signed by a registered medical practitioner or by a coroner after inquest.
>
> That in each sanitary district a registered medical practitioner should be appointed as public medical certifier of the cause of death in cases in which a certificate from a medical practitioner in attendance is not forthcoming.
>
> That a medical practitioner in attendance should be required, before giving a certificate of death, personally to inspect the body.
>
> That a form of a certificate of death should be prescribed, and that in giving a certificate a medical practitioner should be required to use such a form.

The United States Congress, trailing behind the British Parliament, enacted no standard registration act until 1903. Prior to that date, any attempts at standardization were left to the individual states.

In Massachusetts deaths were reported on sheets of paper measuring eighteen by twenty-four inches, with approximately forty records to a page. Information asked for (but not always provided) included date of death, date of record, name of deceased, sex, marital status, age, disease or other cause of death, residence and place of death, occupation, place of birth, and names and birthplaces of parents. These questions are generally to be found on all records of that period. The California form added only a line for the signature of the attending physician or coroner. New Jersey requested the same information but asked for a more detailed explanation of the cause of death. This was optional, however, and the space provided for it was usually left blank. The New York form, the most comprehensive of its time, also required data concerning the burial.

These early death registrations lacked many significant data: cause of death was often omitted, and no questions were asked regarding an autopsy report or the time, place, and manner of the death—relevant information needed for statistical purposes and criminal investigation.

When the United States Census Bureau was making its preparations for the 1880 census, it decided to rely upon registration records instead of mortality enumerations wherever possible, and made a study of state and local forms to determine where these registration records were adequate to its purpose. Wide variation was noted in the ways in which items were worded and data recorded. The study revealed the inherent disadvantages of allowing each state to enact its own registration system without guidelines provided by some central authority (Colby 1965). Therefore, with the aid of the American Public Health Association, the Census Bureau developed what may be termed a model death certificate and prescribed its use by the states, but the use of a standard certificate for the registration of deaths was not approved by Congress, as we have seen, until 1903.

11

THE DEATH CERTIFICATE

As this book is about death, so this chapter is about a unique document that memorializes death. We have seen that death is an epistemologically curious event that cannot be experienced and never has been directly reported. The document of which we speak is one that must, without exception, be completed (in due time) for each reader, and which, under no possible circumstances in this world, can any reader ever see completed for himself. The document is, of course, the death certificate.

That interesting document is much more than just a document. It is better understood as that special form which gives operational meaning to death and which, in fact, defines its current dimensions. It reflects the ways in which man—administrative and forensic man, at any rate—thinks about death and the ways in which he believes it occurs.

The impact of the death certificate is considerable. It holds a mirror to our mores; it reflects some of our deepest taboos; it can directly affect the fate and fortune of a family, touching both its affluence and its mental health; it can enhance or degrade the reputation of the decedent and set its stamp on his postself career. But if the impact of the death certificate is great, its limitations are of equal magnitude. In its present form the death certificate is a badly flawed document.

Today most states follow the format of the U.S. Standard Certificate of Death. Most relevant to our present interests is the item which reads: "Accident, suicide or homicide (specify)". When none of these is checked, a natural mode of death is, of course, implied. Only two states, Delaware and Virginia, have made all four of these modes of death explicit on the death certificate (and have included "undetermined" and "pending" categories as well). Curiously enough, Indiana included these modes of death on the death certificate form from 1955 to 1968, but then revised the form in 1968 and now provides no item for mode of death; nor, surprisingly, does the current Massachusetts death certificate contain an accident-suicide-homicide item.

In addition to the U.S. Standard Certificate, the International Classification of Diseases and Causes of Death plays a major role in determining the way a specific death may be counted—and thus in the apparent change in statistical causes of death from decade to decade. For example, the definitions of suicides and accidents were changed in the Seventh Revision (1955) and Eighth Revision (1966) of the International Classification, and the numbers of suicides and accidents changed along with the definitions. When the Seventh Revision was put into effect for the data year 1958, the death rate for suicides increased markedly over 1957. In part, one can find the explanation in this paragraph (U.S. Department of Health, Education and Welfare 1965):

> About 3.3 percent of the total suicide rate for 1958 as compared with that for 1957 resulted from the *transfer of a number of deaths from accident to suicide.* In 1958 a change was made in the interpretation of injuries where there was some doubt as to whether they were accidentally inflicted or inflicted with suicidal intent. Beginning with the Seventh Revision for data year 1958, "self-inflicted" injuries with no specification as to whether or not they were inflicted with suicidal intent and deaths from injuries, whether or not self-inflicted, with an indication that it is not known whether they were inflicted accidentally or with suicidal intent, are classified as suicides. The change was made on the assumption that the majority of such deaths are properly classified as suicide *because of the reluctance of the certifier to designate a death as suicide unless evidence indicates suicidal intent beyond the shadow of a doubt.* The magnitude of the comparability ratios for suicide varied considerably with means of injury, from 1.02 for suicide by

> firearms and explosives to 1.55 for suicide by jumping from high places. [Emphasis added.]

It would seem that this redefinition led to an apparent 55 percent increase from one year to the next in suicides by jumping from high places. Even more interesting is the official observation that the death certifier would be reluctant to indicate suicide "unless evidence indicates suicidal intent beyond the shadow of a doubt." Clearly the certifier plays an important role in the process of generating mortality data. It is he who makes the subjective judgment of what constitutes conclusive evidence of the decedent's intent. The Eighth Revision (1966), which made the category "Undetermined" available introduced still further problems, apparently shifting many suicidal deaths to the Undetermined category. What is urgently needed is an exploration and description of the current practices of certifying deaths, especially deaths by suicide. We need a uniform system that would eliminate such inconsistencies (or confirm the differential unequivocally) as for example, 10.9 deaths by suicide per 100,000 population for Idaho versus 20.2 for Wyoming. What is required is a "correctional quotient" for each reporting unit–county, state, and nation. Until such information is obtained, available suicidal statistics are highly suspect.

That is what the situation is now. At the turn of the century, an early reference book of the medical science (Abbot 1901) urged reliable death registration bookkeeping:

> The objects secured by a well devised system of death certification are manifold and may be enumerated as follows:
>
> 1. Questions relating to property *rights* are often settled by a single reference to a record of a death.
>
> 2. The official certificate of a death is usually required in each case of claim for *life insurance.*
>
> 3. Death certificates settle many disputed questions in regard to *pensions.*
>
> 4. They are of great value in searching for records of *genealogy.*
>
> 5. A death certificate frequently furnishes valuable aid in the *detection of crime.*
>
> 6. Each individual certificate is a contribution *causa scientiae.* Taken collectively they are of great importance to physicians, and especially to health officers, in the study of

disease, since they furnish valuable information in regard to its causes, its prevalence, and its geographical distribution.

All that is well and good. But that was three-quarters of a century ago. It is time to take full account of the enormous scientific and intellectual developments of the twentieth century, especially the psychiatric revolution that began with Freud and has grown steadily during the past three generations. At least four more functions for the death certificate might be added to the half-dozen listed in 1901:

7. The death certificate should reflect the dual nature of death; that is, its private nature (as it is almost experienced by the decedent) and its public nature (as it is experienced and accounted for by others).

8. It should reflect the type of death that is certified—brain death (a flat electroencephalographic record), somatic death (no respiration, heartbeat, reflexes), or one of the other types discussed in chapter 7.

9. It should include space for the specification of death by legal execution, death in war or military incursions, death by police action, and others of the sort.

10. Perhaps most important, it should abandon the anachronistic Cartesian view of man as a passive biological vessel on which the fates work their will, and instead reflect the contemporary view of man as a psycho-socio-biological organism that can, and in many cases does, play a significant role in hastening its own demise. This means that the death certificate should contain at least one item on the decedent's *intention* vis-à-vis his own death. It is not enough to state that a death was natural, accidental, suicide, or homicide; we should know too whether it was intentioned, subintentioned, or unintentioned.

Let us look at the typical death certificate reproduced here. What is there seems fairly straightforward. The items speak for themselves. But not everything is so obvious. For example, it is possible to think of the items on the death certificate as being divided into three groups: The top third of the certificate has to do with identification of the decedent. It establishes exactly who that person was: name, date of birth, place of birth, spouse's name, mother's maiden name, Social Security number, etc.—items calculated to distinguish one John Allen Smith from any other.

The middle section of the certificate relates to cause or causes of the death. There is some worldwide agreement as to what causes are to be listed. Indeed, as we have seen, there is an international

FORM VS-3

STATE FILE NO.

CERTIFICATE OF DEATH

STATE OF MAINE DEPARTMENT OF HEALTH AND WELFARE

PLACE OF DEATH AND USUAL RESIDENCE

1. **PLACE OF DEATH**
   a. COUNTY
   b. CITY, TOWN, OR LOCATION
   c. LENGTH OF STAY IN 1b
   d. NAME OF HOSPITAL OR INSTITUTION (If not in hospital, give street address)
2. **USUAL RESIDENCE** Where deceased lived. If institution: residence before admission
   a. STATE
   b. COUNTY
   c. CITY, TOWN, OR LOCATION
   d. STREET ADDRESS
   e. IS RESIDENCE ON A FARM? YES ☐ NO ☐

DECEDENT PERSONAL DATA

TYPE OR PRINT NAME

3a. NAME OF DECEASED—First Name | 3b. Middle Name | 3c. Last Name

4. DATE OF DEATH Month Day Year

5. SEX

6. COLOR OR RACE

7. Married ☐ Never Married ☐ Widowed ☐ Divorced ☐

8. DATE OF BIRTH

9. AGE (In years last birthday) | If under 1 year: Mos. Days | If under 24 hrs.: Hrs. Min.

10a. USUAL OCCUPATION (Give kind of work done most of working life, even if retired)

10b. KIND OF BUSINESS OR INDUSTRY

11. BIRTHPLACE (State or foreign country)

12. CITIZEN OF WHAT COUNTRY?

13. FATHER'S NAME

14. MOTHER'S MAIDEN NAME

15. NAME OF SPOUSE (If Married)

16. WAS DECEASED EVER IN U.S. ARMED FORCES? (Yes, no, or unk.) (If yes, give war or dates of service)

17. SOC. SECURITY NO.

18. INFORMANT Address

CAUSE OF DEATH

PLEASE TYPE OR PRINT

19. **CAUSE OF DEATH** (Enter only one cause per line for (a), (b), and (c).)

PART I. DEATH WAS CAUSED BY:

IMMEDIATE CAUSE (a) ____________________

Conditions, if any, which gave rise to above cause (a) stating the underlying cause last.

DUE TO (b) ____________________

DUE TO (c) ____________________

INTERVAL BETWEEN ONSET AND DEATH

PART II. OTHER SIGNIFICANT CONDITIONS contributing to death but not related to the terminal disease condition given in Part I(a)

20. WAS AUTOPSY PERFORMED? YES ☐ NO ☐

DEATH DUE TO EXTERNAL VIOLENCE

21a. ACCIDENT ☐ SUICIDE ☐ HOMICIDE ☐

21b. DESCRIBE HOW INJURY OCCURRED. (Enter nature of injury in Part I or Part II of item 19.)

21c. TIME OF INJURY Hour a.m. p.m. Month, Day, Year

21d. INJURY OCCURRED WHILE AT WORK ☐ NOT WHILE AT WORK ☐

21e. PLACE OF INJURY (e.g., in or about home, farm, factory, street, office bldg., etc.)

21f. CITY, TOWN, OR LOCATION COUNTY STATE

PHYSICIAN'S OR MEDICAL EXAMINER'S CERTIFICATION

22a. **MEDICAL EXAMINER:** I hereby certify that death occurred at the time and from the causes stated above, and that I held an (investigation) (autopsy) on the remains of the deceased as required by law.

22b. **PHYSICIAN:** I hereby certify that I attended the deceased from to and last saw him alive on Death occurred at m. on the date and from the causes stated above.

23a. **SIGNATURE** (Degree or title)

23b. ADDRESS

23c. DATE SIGNED

FUNERAL DIRECTOR AND REGISTRAR

24a. BURIAL, CREMATION, REMOVAL (Specify)

24b. DATE

24c. NAME OF CEMETERY OR CREMATORY

24d. LOCATION (City, town, or county) (State)

25. FUNERAL DIRECTOR ADDRESS

26. DATE RECD. BY LOCAL REG.

27. REGISTRAR'S SIGNATURE

classification of diseases and causes of death: an Abbreviated List of 50 Causes and an Intermediate List of 150 Causes. The International Conference for the Eighth Revision of the International Classification of Diseases, held in Geneva in 1965 (U.S. Department of Health, Education and Welfare 1966), considered compilation of a longer list of 250 to 300 causes.

The bottom third of the death certificate contains a number of items usually related to injury and to such miscellaneous items as place of burial or cremation, name of funeral director or embalmer, and so on—none of which interests us especially. But there is one item of very special interest in this section, usually relative to "violent death," which typically contains only three words: accident, suicide, homicide. The important point is this: If none of these three is checked, it is implied that the death was natural. These four terms, then, represent the four traditionally implied modes of death: natural, accidental, suicidal, and homicidal—what, acronymically, I have called the NASH classification (Shneidman 1963). (The use of modifications and combinations of these terms to yield other labels, such as "probable suicide," "probable accident," "suicide-accident undetermined," and so on, does not change the fact that there are only four main modes of death stated or implied on the present certificate.)

It should be immediately apparent that the cause of death stated on the certificate does not automatically carry with it information as to the specific mode of death. One example should suffice: Asphyxiation due to drowning in a swimming pool does not clearly communicate whether the decedent struggled and drowned (accident), entered the pool with the intention of drowning himself (suicide), or was held under the water until he was drowned (homicide).

In the Western world death is given its administrative dimensions by the death certificate. It is the format and content of this document that determines and reflects the categories in terms of which death is conceptualized and death statistics reported. The ways in which deaths were described and categorized in John Graunt's day and earlier set deep precedents for ways of thinking about death, and they govern our thoughts and gut reactions to death to this day. As we have seen, deaths were then assumed to fall into one of two categories: there were those that were truly adventitious—accidents, visitations of fate or fortune (called natural and accidental)—and there were those that were caused by a culprit who needed to be sought out and punished (called suicidal and homicidal deaths). In the case of suicide, the victim and the assailant were combined in the same person and the offense was designated as a crime against oneself, a *felo de se*. England did not cease to classify suicide as a crime until 1961, and in this country it remains a crime in nine states to this day (Litman 1970).

The importance of the certification of the mode of death—of the coroner's function—can now be seen: it not only set a stamp of innocence or stigma upon the death, but also determined whether the decedent's estate could be claimed by his legal heirs (natural or accidental deaths) or by the crown or local lord (a suicide or a murderer). That was certainly one important practical effect of the death certificate. This bias (in relation to suicide, at any rate) is reflected today in the ways insurance policies are written. I would assert that the NASH categories of death were implied as early as the sixteenth century in English certification, and that this submanifest administrative taxonomy of death has beguiled most men into thinking that that is the way death phenomena really are; which, of course, is not necessarily so at all.

Although it may be platitudinous to say that in each life the inevitability of death is an inexorable fact, there is nothing at all inexorable about our ways of dimensionalizing death. Conceptualizations of death are man-made and mutable; what man can make he can also clarify and change. Indeed, changes in the conceptualizations of death are constantly occurring, notwithstanding the NASH notions of death that have held on for centuries after they became anachronistic. Each generation becomes accustomed to its own notions and thinks that these are universal and ubiquitous.

From the time of John Graunt and his mortuary tables in the seventeenth century through the work of Cullen in the eighteenth century and William Farr in the nineteenth century, the adoption of the Bertillon International List of Causes of Death in 1893, and the International Conference for the Eighth Revision of the International Classification of Diseases as recently as 1965, the classification of causes of death has constantly been broadening in scope, the changes characterized primarily by attempts to reflect additions to knowledge, particularly those contributed by the new professions as they have developed—anesthesiology, pathology, bacteriology, immunology, advances in obstetrics and surgery, and most recently, the behavioral sciences.

The traditional natural-accident-suicide-homicide classification of modes of death is demonstrably insufficient: certain deaths cannot be classified as other than equivocal. This can be true, of course, even when *cause* of death is clearly established. Indeed, in the modern medical examiner-coroner's office, it is a very rare case—given the available skills of pathologist, microscopist, and toxicologist—in which the cause of death cannot be determined. But it does not follow at all that mode of death can be so clearly stated. As we saw earlier, an estimated 10 to 15 percent of all coroners' cases are equivocal as to mode of death, the alternatives usually being accident and suicide.

The most serious fault in the certification of equivocal death is the lack of any attempt to establish the *intention* of the decedent in regard to his own demise. The decedent's intention—not his stomach or lung contents or his brain pathology—is what operationally distinguishes suicide from the other three modes. And the decedent's intention cannot be found in the test tube or under the microscope. Often, however, it can be discovered by conscientious interviewing of people who knew various aspects of his life style and specific behavior immediately prior to his death. A total autopsy ought to include the services of the behavioral scientist—psychologist, psychiatrist, sociologist, social worker. We call this procedure the "psychological autopsy." It will be discussed in some detail in the following chapter.

Much of what I have had to say about the NASH classification of death has impugned its heuristic and scientific usefulness. What, in fact, is its major shortcoming? By far its greatest inadequacy lies in the fact that it emphasizes relatively trivial elements in the death of a human being while omitting altogether the psychological role he may have played in his own demise. The NASH classification, Cartesian and apsychological in spirit, implies that the human being is a biological machine to which things happen, rather than a vital, introspective, unique individual who often unconsciously plays a decisive role in his own fate. In other words, it leaves man out.

I propose that we put him in. We could begin by adding the item I have already suggested—an indication of the decedent's intention regarding his own death. This item might be labeled "Imputed Lethality," since this judgment can only be inferential, and I suggest that it consist of four designations: "High," "Medium," "Low," "Absent."

High imputed lethality would indicate that the decedent definitely wanted to die, and played a direct and conscious role in his own death. The death was due primarily to the decedent's conscious wish to be dead, and to his actions in carrying out that wish either by some recognized means of suicide (jumping from a high place, shooting himself in the head) or by deliberately goading someone to kill him, refusing life-saving procedures, stopping a prescribed medical regimen, or some other act of commission or omission that he knew would result in his death.

Medium imputed lethality would indicate that the decedent played an important role in effecting his own death. His behavior in some degree hastened the event—carelessness, foolhardiness, neglect of self, rash judgment, gambling with death, laxness in following a prescribed life-saving medical regimen, active resignation to death, drug abuse, habitual drunkenness, "tempting fate," "asking for trouble."

Low imputed lethality would indicate that the decedent played some small but not insignificant role in effecting or hastening his own demise. The difference between medium and low imputed lethality is one of degree, not of kind.

When imputed lethality is absent, the decedent has played no role in effecting his own death. The death was due entirely to assault from outside the body (in no way invited by the decedent) or to failure within the body (in a decedent who unambivalently wished to continue to live).

This is a classification that seems to me to be meaningful; it is more fair than the NASH categories alone. At present, individuals of higher social status who commit suicide are more likely to be assigned the mode of accident or natural death than are individuals of low social status whose suicidal intent appeared no less ambiguous. If the term is to have any meaning at all, it should be used fairly across the board, measured by the individual's intention.

Perhaps more important from the larger view, the lethality-intention item would provide an unexampled source of information by means of which biostatisticians, public health officials, and social scientists could assess the mental health of any community. It is obvious that the number of deaths that are caused, hoped for, or hastened by the decedents themselves is a measure of the prevalence of psychological disorder and social stress. At present we do not have this measure, and we need it.

It might be protested, inasmuch as the assessments of these intention states involve the appraisal of unconscious factors, that some workers (especially lay coroners) cannot legitimately be expected to make the kinds of psychological judgments required for this type of classification. But medical examiners and coroners throughout the country are making judgments of precisely this nature every day of the week. When a coroner must evaluate a possible suicide, he acts, perhaps without realizing it, as psychiatrist and psychologist, as both judge and jury: any certification of death as suicide implies some judgment or reconstruction of the victim's motivation or intention. But it would be far better if these psychological dimensions of death were made explicit through use of a lethality-intention scale than to allow them to remain implicit and be used in an influential manner. The dilemma is between the present usable, oversimplified classification on the one hand, and a somewhat more complex but more precise classification on the other.

In Marin County, California, the coroner's office* is currently assessing each death processed by that office in terms of both the traditional NASH classification of mode of death and the lethality intention of the decedent. For a two-year period, 1971-1972 (978 cases), the breakdown was as follows:

(1) Natural deaths (630): high lethality intent, none; medium lethality, 33 (5%); low, 37 (6%); absent, 560 (89%).

(2) Accidental deaths (176): high lethality intent, 2 (1%); medium, 77 (44%); low, 40, (22%); absent, 57 (33%).

(3) Suicidal deaths (131): high lethality intent, 131 (100%).

(4) Homicidal deaths (37): high lethality intent, none; medium, 20 (54%); low, 9 (24%); absent, 8 (22%). Four deaths were of unknown origin.

The first thing we notice is that *some* natural, accidental, and homicidal deaths were classified as having *some* degree of lethal intention. If the medium- and low-intention categories are combined, then over one-fourth (26 percent) of all natural, accidental, and homicidal deaths (216 of 847) were deemed to be subintentioned. If one adds the suicidal deaths (in which the decedent obviously has played a role), then only 64 percent—or 625 of 978—of all deaths were deemed to have been totally adventitious; or, conversely, 36 percent were deemed to have some psychological components.

Also of special interest in these Marin County data is the finding that coroners can, with no more apparent difficulty than they experience in assigning deaths to the NASH categories, simultaneously (and by essentially the same processes of inference and induction) assign deaths to psychological (intentional) categories as well. It is an important pioneer effort that deserves widespread emulation.

*I am especially grateful to Keith C. Craig, coroner's deputy, Marin County, for his interest and help in supplying these data.

12

THE PSYCHOLOGICAL AUTOPSY

As long as deaths are classified solely in terms of the four NASH categories, it is immediately apparent that some deaths will, so to speak, fall between the cracks, and our by now familiar problem of equivocal death will continue to place obstacles in our path to understanding human beings and their dying. Many of these obstacles can be cleared away, as we have seen, by reconstructing, primarily through interviews with the survivors, the role that the deceased played in hastening or effecting his own death. About a decade ago I labeled this procedure the "psychological autopsy." Initially its main purpose was to clarify situations in which the mode of death was not immediately clear. In the last few years, especially with the interesting and valuable work of Litman, et al. (1963) and of Weisman and Kastenbaum (1968), the term "psychological autopsy" has come to have other, slightly different meanings. At present there are at least three distinct questions that the psychological autopsy can help to answer:

1. *Why* did the individual do it? When the mode of death is, by all reasonable measures, clear and unequivocal—suicide, for example—the psychological autopsy can serve to account for the reasons for the act or to discover what led to it. Why did Ernest Hemingway "have to" shoot himself (Hotchner 1966)? Why did

former Secretary of Defense James Forrestal kill himself (Rogow 1963)? We can read a widow's explicit account of how she helped her husband, dying of cancer, cut open his veins in Lael Tucker Wertenbaker's *Death of a Man* (1957). Some people can understand such an act; others cannot. But even those who believe they understand cannot know whether their reasons are the same as those of the cancer victim or his wife. What were *their* reasons? In this type of psychological autopsy, as in the following type, the mode of death is clear, but the reasons for the manner of dying remain puzzling, even mysterious. The psychological autopsy is no less than a reconstruction of the motivations, philosophy, psycho-dynamics, and existential crises of the decedent.

2. *How* did the individual die, and when—that is, why at that particular time? When a death—usually a natural death—is protracted, the individual dying gradually over a period of time, the psychological autopsy helps to illumine the sociopsychological reasons why he died *at that time*. This type of psychological autopsy is illustrated by the following brief case from Weisman and Kastenbaum (1968):

> An 85-year-old man had suffered with chronic bronchitis and emphysema for many years but was alert and active otherwise. He had eagerly anticipated going to his son's home for Thanksgiving, and when the day arrived he was dressed and ready, but no one came for him. He became more concerned as the hours went by. He asked the nurse about messages, but there were none, and he finally realized that he would have to spend the holiday at the hospital. After this disappointment the patient kept more and more to himself, offered little, and accepted only minimal care. Within a few weeks he was dead.

The implication here is that the patient's disappointment and his resignation to it were not unrelated to his sudden downhill course and his death soon afterward; i.e., if his son had come to take him out for Thanksgiving, the old man would have lived considerably longer than he did. This man's death, like some others we considered in Chapter 8—voodoo deaths, unexplained deaths under anesthesia, and "self-fulfilling prophecy" deaths, for example—must be considered subintentioned. There can be little doubt that often some connection exists between the psychology of the individual and the time of his death. (Shneidman 1963).

There is, of course, a wide spectrum of applicability of this concept. When a person has been literally "scared to death" by his belief in the power of voodoo, the role of the victim's psychological state seems fairly obvious; and it is difficult to believe that there was no psychological connection between the fatal stroke of Mrs. Loree Bailey, owner of the Lorraine Motel in Memphis, and the assassination of Martin Luther King, Jr., at the motel three hours earlier. But in many other cases any relationship between the individual's psychological state and the time of his death seems difficult or impossible to establish.

As an example of the problems raised by this concept, consider the following case, reported in *The New York Times* of June 26, 1968:

> *WIDOW, 104, DIES IN COTTAGE*
> *SHE ENTERED AS 1887 BRIDE*
>
> Mrs. John Charles Dalrymple, 104 years old, died here [Randolph Township, N. J.] yesterday in the cottage to which she came as a bride in 1887.
>
> Her husband brought her in a sleigh to the house, which she was to leave next week to make way for the new Morris County Community College. . . .

The main question here, as in Weisman and Kastenbaum's case of the old man who was left alone on Thanksgiving, is: Might even this person have lived at least a little longer had she not suffered the psychologically traumatic threat of being dispossessed from the home where she had lived for eighty-one years? Or does the question in this particular case tax one's commonsense credulity?

3. *What* is the most probable mode of death? This was the question to which the psychological autopsy was initially addressed. When cause of death can be clearly established but mode of death is equivocal, the purpose of the psychological autopsy is to establish the mode of death with as great a degree of accuracy as possible. This original use of the psychological autopsy grew out of the joint efforts of the Los Angeles County Chief Medical Examiner-Coroner, then Dr. Theodore J. Curphey, and the staff of the Los Angeles Suicide Prevention Center to apply the skills of the behavioral

sciences to the problem of equivocal death (Curphey 1961, 1967; Litman, et al. 1963; Shneidman and Farberow 1961). Here are three simplified examples:

Cause of death: asphyxiation due to drowning. A woman found in her swimming pool. Question as to correct mode: Did she "drown" (accident) or was it intentional (suicide)?

Cause of death: multiple crushing injuries. A man found dead at the foot of a tall building. Question as to correct mode: Did he fall (accident) or did he jump (suicide)? Or, even, was he pushed or thrown (homicide)?

Cause of death: barbiturate intoxication due to overdose. A woman found in her bed. Question as to correct mode: Would she be surprised to know that she was dead (accident), or is this what she had planned (suicide)?

The typical coroner's office, whether headed by a medical examiner or by a lay coroner, is more likely to be accurate in its certification of natural and accidental deaths than those deaths that might be suicides. A distinguished pathologist says, "A major reason for this, of course, is that both the pathologist and the lay investigator lack sufficient training in the field of *human behavior* to be able to estimate with any fair degree of accuracy the mental processes of the victim likely to lead to suicidal death. It is here that the social scientists, with their special skills in human behavior, can offer us much valuable assistance," (Curphey 1961).

The professional personnel who constitute a death investigation team obviously should hold no brief for any particular mode of death, such as suicide, over any other. In essence, the members of the death investigation team interview persons who knew the deceased—the spouse, grown children, parents, friends, neighbors, co-workers, physicians, and so on—and attempt to reconstruct his life style. They focus particularly on the decedent's life style just prior to his death. If the information they receive contains any clues pointing to suicide, their especially attuned ears will recognize them. They listen for any overt or covert communications that might illuminate the decedent's role (if any) in his own demise. They then make a reasoned extrapolation of the victim's intention and behavior over the days and minutes preceding his death, using all the information they have obtained.

An outline for the psychological autopsy procedure is presented below, with the caution that the investigator must never forget that he is asking questions that are very painful to people in a grief-laden situation. The person who conducts a psychological autopsy should participate in the anguish of the bereaved, work in the service of the mental health of the survivors, and at the same time quietly obtain information that may throw light on the intention (divined from behavior) of the deceased in regard to his own death. Here, then, are the data to be included in the psychological autopsy (Shneidman 1969):

1. Identifying information for victim (name, age, address, marital status, religious practices, occupation, and other details).
2. Details of the death (including the cause or method and other pertinent details).
3. Brief outline of victim's history (siblings, marriage, medical illnesses, medical treatment, psychotherapy, suicide attempts).
4. Death history of victim's family (suicides, cancer, other fatal illnesses, ages of death, and other details).
5. Description of the personality and life style of the victim.
6. Victim's typical patterns of reaction to stress, emotional upsets, and periods of disequilibrium.
7. Any recent—from last few days to last 12 months—upsets, pressures, tensions, or anticipations of trouble.
8. Role of alcohol or drugs in (a) overall life style of victim and (b) his death.
9. Nature of victim's interpersonal relationships (including physicians).
10. Fantasies, dreams, thoughts, premonitions, or fears of victim relating to death, accident, or suicide.
11. Changes in the victim before death (of habits, hobbies, eating, sexual patterns, and other life routines).
12. Information relating to the "life side" of victim (upswings, successes, plans).
13. Assessment of intention; i.e., role of the victim in his own demise.
14. Rating of lethality.
15. Reactions of informants to victim's death.
16. Comments, special features, etc.

The results of these interviewing procedures are then discussed with the chief medical examiner or coroner. Because it is his responsibility to indicate (or amend) the mode of death, all available psychological information should be included in the total data at his disposal. Since a sizable percentage of deaths are equivocal as to mode precisely because these psychological factors are unknown, medical examiners and coroners throughout the country are robbing themselves of important information when they fail to employ the special skills of the behavioral scientists in cases of equivocal death. The skills of behavioral scientists should be employed in the same way that the skills of biochemists, toxicologists, histologists, microscopists, and other physical scientists are used. The time has long since passed when we could enjoy the luxury of disregarding the basic teachings of twentieth-century psychodynamic psychology and psychiatry. Certification procedures (and the death certificates on which they are recorded) should reflect the role of the decedent in his own demise, and in equivocal cases this cannot be done without a psychological autopsy.

The retrospective analysis of deaths not only serves to increase the accuracy of certification (which is in the best interests of the overall mental health concerns of the community), but also has the heuristic function of providing the serious investigator with clues that he may then use to assess lethal intent in living persons.

And there is still another function that the psychological autopsy serves: in working with the bereaved survivors to elicit data relative to appropriate certification, a skillful and emphathic investigator is able to conduct the interviews in such a way that they are of actual therapeutic value to the survivors. Commenting on this important mental health function of the psychological autopsy, Curphey (1961) has stated:

> The members of the death investigation team, because of their special skills, are alert in their interviews with survivors to evidences of extreme guilt, serious depression, and the need for special help in formulating plans for solving specific problems such as caring for children whose parents committed suicide. Since we noted this phenomenon, the coroner's office has, in some few cases, referred distraught survivors of suicide victims to members of the team specifically for supportive interviews even when the suicidal mode of death was not in doubt.

This therapeutic work with the survivor-victims of a dire event is called "postvention" (Shneidman 1967, 1970a), and has been presented in some detail in Chapter 3.

It would seem appropriate to conclude this chapter on the psychological autopsy by presenting, by way of example, some excerpts from two singularly interesting cases.

In Chapters 1 and 9 we touched upon the tragic death of the twenty-eight-year-old man who died in a fire in his apartment the day after the death, from cancer, of his mother. Reprinted below, with changes to assure anonymity, are excerpts of the psychological autopsy of his death. The first necessary step was that the coroner recognized the death as equivocal and referred the case to a death investigation team of social scientists for further inquiry. The following persons were individually interviewed: two personal friends of the deceased, the owner of the decedent's apartment, a friend of the decedent's mother, and a battalion chief of the fire department's arson squad.

"The victim was found dead in his burning apartment. He was burned over 80 percent of his body. Blood analysis showed Ethanol absent; 0.6 mg. percent Secobarbital, and .009 mg. percent Amphetamines. 85% carbon monoxide saturation was found in the blood.

"The firemen had difficulty entering the deceased's apartment because the place was in such a mess. There was debris against the door that had to be moved in order to open the door. There was no furniture against the door. It looked as though someone had thrown things around in a fit of despair. The arson investigation from the Fire Department came up with two possibilities as to the source of ignition: (1) An electrical short from the cord of a small radio at the front of the couch. The radio was still plugged in. The cord ran around a stack of combustible material piled on the floor at the end of the couch. (2) Newspapers placed on the floor over the outlet of a heater. However, the thermostat was found in an off position. It had fallen off the wall. The evidence and arson investigation found that the fire was not intentionally set. Photographs taken by the Police and Fire Departments showed that the body of the victim was headed toward the door. The Battalion Chief stated the fire probably had smoldered for a long time before actually igniting into flames. Thus, it is probable that the victim had a large quantity of carbon monoxide in his body before he might have been aware of the fire, if he were barely conscious.

"The deceased was a 28-year-old unemployed male, who lived alone. The day before his death, his mother died after a lengthy illness with cancer. He had been unemployed for the past few years and was supported entirely by his mother. In recent months, the deceased became more withdrawn and socially isolated, spending most of the time alone in the apartment, with the exception of daily visits to his dying mother in the hospital. During this period the deceased's long-term love relationships with others were deteriorating.

"One of the friends interviewed stated that he had known the deceased for approximately ten years and that he wanted to help him straighten out his life and introduced him to metaphysics. He described the deceased as an avid reader of Austen and Dostoyevsky, a lover of music, and that he possessed an I.Q. of over 180. He also had the ability to read a page once and repeat verbatim what he had read.

"The friend also stated that the deceased's mother was difficult to please and while in the hospital would make demands on her son, i.e. wanting pudding with a special sauce which he would go out and get, then when he brought it to her she no longer wanted it. The deceased visited his mother every day in the hospital. The son was tied to his mother and also dependent upon him (the friend).

"The deceased did not drink with the exception of Pernod on occasion. He had, at one point, taken some Dexamyl pills, but was not taking pills at the present time. The deceased smoked incessantly.

"The deceased called his friend when the mother died. He went to the hospital with him and assisted in the arrangements for burying the mother in another state. The deceased told him at that time his own wish to be cremated in accordance with his metaphysical beliefs. After the arrangement had been completed at the mortuary, they had breakfast together. The deceased was in good spirits, apparently relieved that the long burden of his mother's illness had been lifted from him. He wanted his friend to accompany him home, but he didn't think it was wise since he was not a family member. At two P.M. they both went to his apartment. The apartment was in disarray, the first time the friend had seen the apartment messy; the deceased was usually a fastidious housekeeper.

"With the death and loss of his mother, the deceased attempted that day to re-establish his relationship with his male friend, who was now involved with someone else. In desperation, he went to his friend's apartment and pounded on the door, creating such a disturbance that the police were called and he was arrested, jailed, and released on bail that night.

"In the late afternoon the friend went again to the apartment. The door was bolted and it took deceased five minutes to get from his bed to the door and unlock it. He was groggy and then collapsed. He put him back in bed. He was crying. There was a bottle of secobarbital which had been prescribed to deceased's mother. A few pills were missing. He emptied the rest of the contents and took them. He told the deceased he would return later and that he should get some sleep so that he would be in shape to accompany his mother's body home the next day. He left the deceased in bed sleeping around 4:00 P.M. The door was left unlocked. The deceased had not talked of suicide. When the friend returned about 9:00 P.M. he found the fire department and ambulance in front of the apartment. The door and screen were both locked when the fire department arrived.

"Another friend reported that at 11:00 P.M. on the day the mother died, the deceased called the friend, asking him to please bail him out of jail. He was surprised to receive the call, but the deceased stated, 'I have no one else to call.' He went to the jail and bailed him out. The deceased explained his arrest as being because he had gone to visit a friend of his and was giving him trouble. The man's mother called the police. They did not arrest him for disturbing the peace, but for past traffic tickets which he had accrued and not paid. This friend drove him to his car. The deceased at that time stated, 'I am going to change. I have lost the only person I loved. I must do something better for myself.'

"They went back to the deceased's apartment. The deceased had told him he had taken a few seconal pills that day. He was drowsy. He came home and took a couple of glasses of wine. He allayed his fears about mixing pills and liquor by saying he had taken the pills three hours previously and they no longer were effective. Suddenly, the deceased appeared extremely drunk. He was staggering and wobbling. He wrote a check to the friend for the amount of the bail, but it was unaccepted because it was illegible. Then the deceased fell flat on his face near the coffee table. He picked him up and put him on the bed.

"The next day the deceased was to go to take clothes to the mortuary to be worn by his mother. At 10:30 A.M., this friend called the deceased to remind him. The deceased was very incoherent and drunk sounding. He called again at 2:00 P.M. to check on the deceased's taking the clothes. The deceased responded, 'Stop bothering me,' and hung up. He felt this response was not like him. However, the friend reported that there were times when his mother would tell him of her displeasure with his private life, and he would respond by not talking to her or not visiting her for weeks at a time. His mother had been very worried about her son. For the past three years he was withdrawn, depressed and often would not go out, wouldn't work, would sleep for days and would lock himself in his apartment. His mother tried to get him involved in psychotherapy. He would not go. Although he could not substantiate this, the friend thinks the deceased had tried to kill himself once with a drug overdose. He knows that the mother had been worried that her son might kill himself, that he was a disturbed man and felt lost.

"The owner of the building deceased lived in stated that he spoke to the fire department as to the cause of the fire and they had said that the source was a radio thrown on top of a pile of papers producing a short in the wire of the radio and catching the papers on fire. Everything was thrown around in the apartment. The coffee table, T.V. and other furniture appeared to have been thrown as though deceased had had a tantrum. The deceased was known as a very quiet man. Neighbors didn't know he was there. He had occupied the apartment for several months and sent his check in by mail. The landlord saw him only three times. At one point he was planning to move and with his mother travel leisurely. He was selling his furniture and some of his mother's furniture he had in his apartment. He changed his mind, however, and didn't leave.

"A friend of the decedent's mother was interviewed. She met the deceased several years ago while taking an English class. At that time she states deceased was a bright, nice boy, who was fun to be with. Over the past several years deceased's personality changed and he became very introverted, a recluse, an introvert with a drug habit that became progressively worse and he was constantly taking amphetamines. The deceased first started with Dexamyl, and three months before his death had purchased 4,000 Methedrine tablets from someone she knows. To her knowledge, these pills had all been used at the time of his death.

"She stated that the deceased had not worked for years and although she felt he had the potential, he was never able to mobilize himself into constructive action. About four or five months before his death, the deceased's attitude changed. He began staying in his house all the time and he felt persecuted, had bizarre ideas as to his future plans, i.e., talked about going to graduate school, moving somewhere else and starting over, but was irrational in his ideas. During these months deceased was rejecting overtures of friendship. She would sometimes go over to his apartment and pound on the door. Later he would call her on the phone and accuse her of rejecting him.

"She reported that she felt the deceased's mother put great pressure on him. She hated the fact that he was unmarried and that he had no job, but was also over-protective of him. When the deceased went to the hospital to visit his mother she would sometime have him pluck her eyebrows. On one occasion he brought her flowers on her birthday and the next day when he went to visit her the flowers were not in the room. This disturbed him greatly. He felt nothing he could do was good enough where his mother was concerned, and she was always rejecting his efforts. While she was hospitalized, she needed a colostomy. She called deceased and told him to put his pants on and come down to the hospital. He had to make the decision as to whether or not she should have the colostomy. She—this informant—called the deceased his mother's 'pseudo-husband.'

"About four days before his death, she saw the deceased. He appeared timid, was stammering in his speech and just sat in the chair without talking much. The day he died deceased called her on the telephone. She had never heard him sound so upset. He was hysterical, frightened and desperate. He told her he had no one to turn to and that he had never felt like this before. He seemed to be 'exploding' and begged her and her husband to come over. The deceased told her that the previous night he had taken several seconals and could not sleep. He stated he was disturbed about his relationship with his male friend, went to his house and pounded on the door begging to be let in but the police were called and he was taken to jail.

"She went to his apartment but when she arrived there she found the roommate of his ex-lover there. The deceased appeared calmer than he had been on the telephone. He had had nothing alcoholic to drink but admittd to taking eight or nine seconals. He appeared incoherent and his hands were shaking so that he couldn't hold his cup of coffee. Deceased appeared so 'bedraggled' that she remarked he was acting like Marilyn Monroe and Judy Garland, on uppers and downers—so 'untogether.' The deceased responded by saying,

'Yes, I'm just like them.' She states he was identifying himself with the Monroe-Garland deaths.

"She believes the deceased was in a highly self-destructive life pattern. He was doing everything to make his death come about and she couldn't imagine how much longer he could live under these conditions. The deceased was 'worn down,' having suffered the rejection and death of his mother and the rejection of his lover. He had nothing in his life. She believes that the deceased's mother didn't want him to live after she died—that she wanted to take him with her. She believes the deceased shared his mother's casket on the trip to be buried.

"In summary, we find a dependent, unemployed, unmarried male, who, upon the death of his mother, and upon the awareness that his boyfriend formed another love relationship, found himself abandoned and alone. Although he appeared suicidal, he may have died when his apartment accidentally caught fire. Evidence at the scene suggests that the deceased apparently had a tantrum-like episode which consisted of his throwing things around his apartment, following which he, very likely, fell into an exhausted deep sleep. Further, he had ingested some secobarbital and amphetamines before the fire started.

"On the basis of the investigation—both the psychological and physical aspects of this case—it is recommended that the mode of death be certified as a Probable Accident."*

The second illustration—quite different from the previous one—is reprinted from a uniquely comprehensive study of death and lives by Herman Melville.† It is the case of the equivocal death—was it accident, suicide, or what?—of Melville's tortured, obsessed, fury-

*A few editorial comments: First I believe that, on the basis of all the available evidence, the category "Accident-Suicide/Undetermined" would have represented a more accurate recommendation for certification. But my further reaction to this tragic death—in addition to anguishing personally over the fact that his cries for help had not been answered—was to eschew the confines of the NASH categories altogether and to conceptualize the death in my own mind as a subintentioned death, i.e., a death that possessed both adventitious and psychologically laden (death-oriented) elements which concatenated at that moment in his life to produce his untimely demise.

†This psychological autopsy of Captain Ahab appeared as part of an essay by me entitled "Orientations Toward Death" in the book *The Study of Lives* (1963), a volume done in honor of Henry Murray by several of his colleagues and former students, and is reprinted with the kind permission of the Atherton Press.

driven man in a deathful vendetta with Moby Dick: Captain Ahab of the *Pequod.**

The procedure called the psychological autopsy involves obtaining psychological data about the behavior and statements of the deceased in the days before his death, from which an extrapolation of intention is made over the moments of, and the moments directly preceding, his cessation. In the case of Captain Ahab, I shall proceed as though I were preparing a report for an imaginary Nantucket coroner, including some sort of recommendation as to what label would be most appropriate on Ahab's death certificate. The focus will be to attempt to come to some kind of resolution concerning Ahab's intention types. But first, some facts: Specifically how did the end of his life occur?

For Ahab's death, we have the following account (from Chapter 135) of his last actions: "The harpoon was darted; the stricken whale flew forward; with igniting velocity the line ran through the groove;—ran foul. Ahab stooped to clear it; he did clear it; but the flying turn caught him round the neck, and noiselessly as Turkish mutes bowstring their victim, he was shot out of the boat, ere the crew knew he was gone." On first thought, it might sound as though Ahab's death were pure accident, an unintentioned death, the cessation of a death-postponer; but let us see where our second thoughts lead us. Perhaps there is more.

It is possible to view *Moby Dick* as a great, sonorous Mahlerlike symphony—*Das Lied von der See*—not primarily about the joy of life nor the pessimism engendered by a crushing fate, but rather as a dramatic and poetic explication of the psychodynamics of death. And, within the context of this thought, is it not possible that Moby Dick, the great white whale, represents the punishment of death itself? In Chapter 28, when Ahab makes his first appearance on the *Pequod* at sea, the word 'white' is used three times in one paragraph to describe Ahab: a head-to-toe scar on Ahab's body, 'lividly whitish;'

*The inspiration for my analysis of Captain Ahab came entirely from the masterful psychological study of *Moby Dick* by Henry Murray, particularly his essay "In Nomine Diaboli," originally printed in *New England Quarterly,* 24 (1951), and since reprinted in *Moby Dick Centennial Essays* (Dallas: Southern Methodist University Press, 1953); in Milton R. Stern (ed.), *Discussions of Moby Dick* (Boston: D. C. Heath, 1960); in Richard Chase (ed.), *Melville: A Collection of Critical Essays* (Englewood Cliffs, N. J.: Prentice-Hall, 1962); and in *Psychology Today* (September 1968).

an allusion to a 'white sailor,' in the context of Captain Ahab's being laid out for burial; and 'the barbaric white leg upon which he partly stood.' Everywhere, reference to the pallor of death; and if there is still any question, the case for 'white death' is made explicit in the discussion of the whiteness of the whale (Chapter 42), in which we are told:

> It cannot well be doubted, that the one visible quality in the aspect of the dead which most appalls the gazer, is the marble pallor lingering there; as if indeed that pallor were as much like the badge of consternation in the other world, as of mortal trepidation here. And from that pallor of the dead, we borrow the expressive hue of the shroud in which we wrap them. Nor even in our superstitions do we fail to throw the same snowy mantel round our phantoms; all ghosts rising in a milk-white fog—Yea, while these terrors seize us, let us add, that even the king of terrors, when personified by the evangelist, rides on his pallid horse.

And if the great white whale is death, then is not the sea itself the vessel of death? Melville sets this tone for his entire heroic narrative in his stunning opening passage:

> Call me Ishmael. Some years ago—never mind how long precisely—having little or no money in my purse, and nothing particular to interest me on shore, I thought I would sail about a little and see the watery part of the world: It is a way of driving off the spleen, and regulating the circulation. Whenever I find myself growing grim about the mouth; whenever it is a damp, drizzly November in my soul; whenever I find myself involuntarily pausing before coffin warehouses, and bringing up the rear of every funeral I meet; and especially whenever my hypos get such an upper hand of me, that it requires a strong moral principle to prevent me from deliberately stepping into the street, and methodically knocking people's hats off—then, I account it high time to get to sea as soon as I can. This is my substitute for pistol and ball. With a philosophical flourish Cato throws himself upon his sword; I quietly take to the ship.

If any case is to be made for subintention then, at the least, two precipitating forces need be explored: unconscious motivation and ambivalence. Ahab's chronicler would not have, in principle, resisted the concept of subintention on the grounds of its involving unconscious motivation, for (in Chapter 41) he says:

> Such a crew, so officered, seemed specially picked and packed by some infernal fatality to help him to his monomaniac revenge. How it was that they so aboundingly responded to the old man's ire—by what evil magic their souls were possessed, that at times his hate seemed almost theirs; the White Whale as much their insufferable foe as his; how all this came to be—what the White Whale was to them, or how to their unconscious understandings, also, in some dim, unsuspected way, he might have seemed the gliding great demon of the seas of life—all this to explain, would be to dive deeper than Ishmael can go. The subterranean miner that works in us all, how can one tell whither leads his shaft by the ever shifting, muffled sound of his pick?

That which is most sharply and most accurately characteristic of the subintentioned person—namely, the ubiquitous ambivalence, the pervasive psychological coexistence of logical incompatibles—is seen vividly in the force and counterforce, the internal dialogue of life and death within Ahab (Chapter 51):

> Walking the deck with quick, side-lunging strides, Ahab commanded the t'gallant sails and royals to be set, and every stunsail spread. The best man in the ship must take the helm. Then, with every mast-head manned, the piled-up craft rolled down before the wind. The strange, upheaving, lifting tendency of the taff-rail breeze filling the hollows of so many sails, made the buoyant, hovering deck to feel like air beneath the feet; while still she rushed along, as if two antagonistic influences were struggling in her—one to mount directly to heaven, the other to drive yawningly to some horizontal goal. And had you watched Ahab's face that night, you would have thought that in him also two different things were warring. While his one live leg made lively echoes along the deck, every stroke of his

dead limb sounded like a coffin-tap. On life and death this old man walked.

In any psychological autopsy it is important to examine the method or the instrument of death to discover, if possible, its presumed lethality in *the victim's mind.* Ahab was garroted by a free-swinging whale line. We are warned (in Chapter 60) that 'the least tangle or kink in the coiling would, in running out, infallibly take somebody's arm, leg, or entire body off'; we are forewarned 'of this man or that man being taken out of the boat by the line, and lost;' and we are warned again, 'All men live enveloped in whale lines. All are born with halters round their necks; but it is only when caught in the swift, sudden turn of death, that mortals realize the silent, subtle, ever-present perils of life.' Ahab knew all this, nor was he a careless, accident-prone man. The apothecary knows his deadly drugs, the sportsman knows the danger of his weapons; the whaler captain—that very whaler captain who, instead of remaining on his quarterdeck, jumped to 'the active perils of the chase' in a whaleboat manned by his 'smuggled on board' crew—ought to know his whale lines.

I would now pose some questions concerning Ahab's demise: Was his death more than simple accident? Was there more intention that unintention? Was his orientation in relation to death entirely that of death-postponing? Are there discernible subsurface psychological currents that can be fathomed and charted, and is there related information that can be dredged and brought to the surface? Specifically, can Ahab's death be described as victim-precipitated homicide; that is, is this an instance in which the victim stands up to subjectively calculated overwhelming odds, inviting destruction by the other? Let us see.

Ahab led a fairly well-documented existence, especially insofar as the dark side of his life was concerned. *Moby Dick* abounds with references to various funereal topics: sleep, coffins, burials, soul, life-after-death, suicide, cemeteries, death, and rebirth. But in a psychological autopsy we are primarily interested in interview data from everyone who had known the deceased, especially in what our informants can tell us about Ahab's personality insofar as his orientations toward death are concerned. In some important ways Captain Ahab's psychological autopsy will be a truncated and atypical one with respect to the range of informants: there is no information from spouse, parents, progeny, siblings, collaterals, neighbors; there are only mates, some of the more articulate shipboard subordinates, captains of ships met at sea, and, with terrifying biblical certitude, Elijah.

As we know, all the possible informants save Ishmael perished with Captain Ahab and are technically not available for interview. Only Ishmael's observations are direct; all else is secondhand through Ishmael colored by Ishmael, and perhaps with no more veridicality than Plato's reports of Socrates. We shall have to trust Ishmael to be an accurate and perceptive reporter.

Our primary informant, Ishmael, reflected about Captain Ahab in twenty-five separate chapters. Starbuck, the chief mate of the *Pequod,* is next: there are nine separate encounters with, or reports about, his captain. Next is Stubb, the second mate, with seven separate anecdotes. All the others are represented by one or two bits of information apiece: Elijah; Gabriel of the *Jeroboam;* Bunger, the ship's surgeon of the *Samuel Enderby;* the blacksmith; the captain of the *Bachelor;* Flask, the third mate; the Manxman; and the carpenter.

The limitations of space simply do not permit me to document the essence of each informant's remarks, either with appropriate quotations or abbreviated résumés. How can I summarize all the data? Perhaps my best course would be to concentrate on the general features that one would look for in any psychological autopsy. Thus, the information distilled from interviews with Ishmael, Starbuck, Stubb, and all the others might, in a dialogue of questions and answers, take the following form:

1. Hidden psychosis? Not at the beginning of the voyage, but certainly at the end (and indeed from Chapter 36 on: 'The chick that's in him picks the shell. Twill soon be out.') The madness in Ahab was blatant, open, known. His monomania was the official creed of his ship. Along with his other symptoms, his psychiatric syndrome was crowned with a paranoid fixation. But what matters in Ahab is not so much the bizarrely shaped psychological iceberg which many saw above the surface, but rather the huge subsurface mass of other-destruction and self-destruction. We know the poems about fire and ice. Oxymoronically, Ahab is a torrid, burning, fiery iceberg.

2. Disguised depression? Ahab was openly morbid and downcast. His was not exactly psychotic depression, nor can we call it reactive

depression, for it transcended the bounds of that definition. Perhaps best it might be called a 'character depression,' in that it infused his brain like the let-go blood from a series of small strokes in the hemisphere.

3. Talk of death? The morbid talk of death and killing runs through reports about Ahab like an *idée fixe*.

4. Previous suicide attempts? None is reported.

5. Disposition of belongings? Ahab, after forty solitary years at sea, had little in the way of material possessions. His wife, he said, was already a widow; his interest in the possible profits from the voyage was nil; his withdrawal from meaningful possessions is perhaps best indicated by his flinging his pipe into the sea and dashing his quadrant to the deck—both rash acts for a sea-captain.

In Ahab's conscious mind, he wanted to kill—but have we not said that self-destruction can be other-destruction in the 180th degree? Figuratively speaking, the barb of the harpoon was pointed toward him; his brain thought a thrust, but his arm executed a retroflex. Was his death an 'accident'? If he had survived his psychodynamically freighted voyage and had returned unharmed to Nantucket's pier, *that* would have been true accident. Men can die for nothing—most men do; but some few big-jointed men can give their lives for an internalized something: Ahab would not have missed this opportunity for the world.

What further evidence can be cited bearing on the issue of subintentioned death? With his three harpooners before him with their harpoons turned up like goblets, Ahab (in Chapter 36) commands them, in this maritime immolation scene: "'Drink, ye harpooners! drink and swear, ye men that man the deathful whaleboat's bow—Death to Moby Dick! God hunt us all, if we do not hunt Moby Dick to his death!'" Kill or be killed; punish or be retributed; murder or suicide—how the two are intertwined.

In Ahab's case, we have no suicide note or other holograph of death, but, *mirabile dictu,* we do have (in Chapter 135) Ahab's last thoughts:

> I turn my body from the sun. . . . Oh, lonely death on lonely life! Oh, now I feel my topmost grief. Ho, Ho! from all your furtherest bounds, pour ye now in, ye bold billows of my whole foregone life, and top this one piled comber of my death! To-

> wards thee I roll, thou all-destroying but unconquering whale; to the last I grapple with thee; from hell's heart I stab at thee; for hate's sake I spit my last breath at thee. Sink all coffins and all hearses to one common pool! and since neither can be mine, let me now tow to pieces, while still chasing thee, though tied to thee, thou damned whale! Thus, I give up my spear!

What is to be particularly noted here is Ahab's prescience. 'I spit my last breath at thee,' he says. How does he know that it is to be his last breath? Where are the sources of his premonitions? What are the contents of his subintentions? Does this not remind us of Radney, the chief mate of the *Town-Ho* (Chapter 54), who behaved as if he 'sought to run more than half way to meet his doom'? Is this not exactly what the tantalizer says to his 'all-destroying but unconquering' executioner in cases of victim-precipitated homicide?

It is suggested that Captain Ahab's demise was goal-seeking behavior: his obsessed life was relatively unimportant to him compared with the great press for the discharge of his monomania of hate. He dared, and made, that murderous death-white whale kill him. He could not rest until he was so taken. (Did Satan provoke God into banishing him?) Ahab invited cessation by the risks that he ran; he was a death-chancer. He permitted suicide. Consider Ahab's psychological position: What could he have done, to what purpose would any further voyages have been, if he had killed the symbol of his search? It was, from Ahab's point of view, the time; and in his unconscious wish, it was the "appropriate death."

5

DIMENSIONS OF DEATH

13

PARTIAL DEATH

Death is the most final and most complete ending in life, but it is not the only one. There are many alterations in our lives that we can appropriately refer to as *endings*—conclusions of phases or aspects of our lives, the closing of episodes that irreversibly put a stop to habitually expected stimuli, psychological states, interpersonal relationships, and living patterns.* Such endings short of death—what C. M. Parkes (1971) calls "psycho-social transitions"—often involve mourning and grief as intense as the mourning caused by death itself, and as appropriate. These endings are partial deaths, psychological deaths, or social deaths. Usually we survive them—as, in a special sense, we survive all forms of death.

*The dying process itself (in, for example, cancer or leukemia) often seems, psychologically, to consist of a series of endings, in which the dying person successively, in the midst of incapacitating pain and discomfort, comes reluctantly to face increasingly constricting truncations and losses of his actual and potential competencies, in a lugubrious series of little deaths.

PARTIAL DEATH

We can—we must—become habituated to the constant losses that accompany the inevitable changes through which each life passes: graduating from school (marked by "commencement"), being discharged from the service, losing a job or a lover, having our property destroyed or stolen, completing a creative effort. Most of these losses are more or less quickly replaced: we go to graduate school or to work, we find a new job or a new lover, we file a claim with our insurance company and replace our lost property, we start work on a new book or a new bookshelf. Other losses are more difficult or impossible to replace: the lost of a limb or of one's sight, the death or divorce of a spouse, the death of a child. In a heart-warming account of a musical idiot savant, Viscott (1970) discusses the "little deaths" of creativity:

> There seems to be a unique psychology in the creative person. The customary issues of mastery are present, but have special meaning. During treatment of a creative talent one becomes aware of a sense of repetition of the life cycle in the birth of every new work and a sense of loss and mourning appearing as each is completed, and given up. This depression stems in part from the artist giving form to his deepest feelings. Once the work embodying these feelings is no longer private, is finished, exposed for all to see, it must live or die on its own. The artist is left to continue coping with his own death by creating works which he hopes will be immortal. For one brief moment, the moment of creation, he reaches back, perhaps attains an inchoate sense of unity with a mothering figure, feels complete, less lonely. It is a fleeting moment, beautiful yet terribly sad, making the infinite appear finite, creating a microcosm, and just as quickly it is gone.

When Dean Acheson left his position as Secretary of State and returned to private life (in 1953) he commented: "To leave positions of great responsibility and authority is to die a little."

Parkes, who has done much recent work on the effects of loss by death, has this to say about the effects of the loss of a limb (1971):

> An amputee loses a part of his visible body as well as the functions which previously he could perform by means of the body part. In any particular situation one rather than the other is likely to be seen as the predominant loss and a cause for grief. There are many points of similarity between the loss of a limb and the loss of a spouse, although in some ways the two situations are not comparable. Thus the immediate reaction to amputation is often a state of numbness in which physical and mental feeling is blunted; this is soon succeeded by a phase of distress in which episodes of severe anxiety occur with pining for the lost limb or those functions which are thought to have been lost. The amputee is preoccupied with fears that he will be unable to work or to walk, but at the same time he tries to avoid reminders and prefers his friends not to mention his loss. Depression, restlessness, tension, difficulty in concentration, insomnia, loss of appetite and loss of weight are common symptoms, and the amputee tends to be irritable and to go over the events leading up to the loss in an attempt to find someone or something whom he can blame.

In a previous chapter we explored the notion of subintentioned death. It is important to note that the role of subintention is not at all limited to dying behaviors. Indeed, the individual's unconscious needs, conflicts, and thrusts, as we all know, play important roles not only in his death style, but in his life style as well—the ways in which he elevates, ennobles, expands, or actualizes his life, or diminishes, narrows, or truncates aspects of his inner and outer lives, bringing about the partial death of his psychic or social self. And there are those regrettable times when it hardly seems to matter; when no matter what we do, we die a little.

Henry Murray (1967) has said that when partial death occurs—death of an aspect of the self—

> It is as if the person's primal springs of vitality had dried up, as if he were empty or hollow at the very core of his being. There is a striking absence of anything but the most perfunctory and superficial social interactions; output as well as intake is at a minimum. The person is a nonconductor. To him the human species is wholly uninviting and unlovable, a monotonous round

of unnecessary duplicates; and since everything he sees and every alternative opportunity for action seems equally valueless and meaningless, he has no basis for any choice. In fact, to make even a small decision and to execute it calls for an exhausting effort. Sometimes, he unresistingly and automatically falls in with somebody else's decision; but he is more likely to respond to suggestions with a blanket No, keeping his thoughts hidden from others behind a deaf-and-dumb reserve, the impenetrable wall of a self-made prison.

I was thinking particularly of Melville's forty-year withdrawal from his society—the "Great Refusal" as Weaver called it—and of a patient of mine who resembled Melville in this respect but whose cessation of affect was more total, suffering as he was from what we used to call the "feeling of unreality." His sensations and perceptions of nature and of people he encountered were unusually acute and vivid, but he did not experience other persons as animate beings: they resembled puppets, automatons, mechanical contrivances without any feeling or desires to which anyone could appeal. He saw eyes that were as bright as the glass eyes of a manufactured doll, but he received no intimations of a soul, or consciousness, behind these eyes. Primitive people and children spontaneously animate the inanimate—see a man in the moon who follows them on their walks, as Piaget has described; but here was a man who reversed the process; he inanimated the animate. All empathy was dead in him; he was inert as a stone, unmoved by any of the events or confrontations which moved others.

I have been talking about a diminution or cessation of feeling, one component of consciousness, on the assumption that this condition is somewhat analogous to a cessation of the whole of consciousness. If the cessation of feeling is temporary it resembles sleep; if it is permanent (a virtual atrophy of emotional life) it resembles death, the condition of the brain and body after the home fires of metabolism in the cortex have gone out. In a feelingless state the home fires are still burning but without glow or warmth.

This cessation of feeling is commonly accompanied by withdrawal from one's associates, by grave social refusal. The partially dead repudiate their society, shun the company of others, cut them dead;

and society in turn shuns them. Partial death is thus a form of death-in-life.

PSYCHOLOGICAL DEATH

Psychological death, as distinct from partial death, is a special state in which the individual ceases to be aware of his own existence (Kalish 1968, Knutson 1970). This process can occur well before clinical death, although in most instances it accompanies it. Kalish explains:

> At the end of the life span, the individual, often elderly, ceases to be aware of his self. In most instances this occurs concurrently with physical death; in only a relatively few cases is the terminal patient fully comatose, completely drugged or hopelessly senile. A recent study by Kastenbaum indicated that under 5 percent of the dying geriatric patients were consistently and completely out of contact, even during the few days preceding physical death; about half were in full contact, and the other half were in either partial or irregular contact. There is, obviously, a continuum of psychological death at any given time, and, second, in the proportion of an extended period of time during which total or near-total psychological death is exhibited. Both dimensions are partially the result of environmental interaction which diminishes as the individual's capacities diminish and as the available figures with whom he can interact diminish.
>
> To the extent that a person is psychologically dead, he not only does not know *who* he is—he does not know *that* he is. This can obviously occur well before clinical death, yet the resemblance of psychological death to clinical death is very disturbing to both family and friends and to medical people. Many people react to psychological death, when extensive and apparently irreversible, by suggesting that the patient be allowed to die physically as well; some encourage euthanasia, even though the person is not necessarily suffering discomfort. In our pragmatic, achievement-oriented nation, a man who does not behave like a human being therefore might as well be dead, and therefore is dead.

SOCIAL DEATH

Social death includes phenomena of interest to a number of current investigators (Kalish 1968, Sudnow 1970, Crane 1970). It can be perceived by the self and by others. The most frequent use of the term occurs in connection with the terminally ill or dying patient. The victim considers himself "as good as dead" and literally withers away until he actually is dead. These feelings can usually be traced to an outside source or external cues, such as a terminal diagnosis, which leaves him feeling helpless, hopeless, and isolated. In this case, it is the perception of others that causes the patient to perceive his own social death.

Sudnow (1970) has defined social death as

> a phenomenon incorporating the process of mutual disengagement and rejection by which "organization man"—more precisely, the human being as a member of society—seems prone to take his leave from the land of the living.
>
> This view of the termination of life concerns itself not with how the biological organism expresses life, or lets go of it finally, but with how the organization deals with this personal tragedy. A human institution, or social organization, acquires an integrity of its own, of course. It has a collection of purposes and a need to do its work and maintain itself; this need is in part consistent, but not necessarily coincidental, with the purposes and needs of individual members of society. Recognizing this much, we may speculate that the death of an individual is of about the same significance to the social organization as the death of a cell is to a biological organism. It happens all the time. Ordinarily and in small numbers, it is not considered of great moment. Other life goes on, unaffected or affected.
>
> The cell analogy may deeply offend our sense of values, but it is useful in pointing out that death is an inevitable characteristic of life; from a logical standpoint, man is in the process of dying throughout life. We choose to overlook this technicality, and much prefer to procrastinate in accepting its ultimate reality. Thus, the statement, "He is dying," or in hospital jargon, "The patient is terminal," is acceptable for application only under a rather restricted set of circumstances. These circumstances are in essence fairly elementary. There comes a time when the seriously ill patient is not holding his own or getting better,

> but appears to be declining. In view of the patient's condition and his failure to respond to treatment, the physician gives up hope for recovery. He reaches the judgment that there is nothing further he can do; he may put some time limit, usually a short term, on the patient's survival.
>
> At this point, the patient becomes a candidate for social death, in a prelude to biological death. Social death begins when the institution, accepting impending death, loses its interest or concern for the dying individual as a human being and treats him as a body—that is, as if he were already dead.

Within the context of everyday living there are other kinds of social death: a change in the living environment, for example, or separation from others with whom one's life has been closely integrated, can be perceived as social death by the self, by others, or by both. We are all socially dead to someone.

Civil death is defined as that separation between society and the individual which incurs the loss of the individual's civil rights. As the concept was originally applied in England, one was declared civilly dead if he entered a monastery, was found guilty of a felony, or renounced the realm. In later years the first two requirements were amended. In the United States conviction for a felony deprives an individual of certain civil rights, such as voting privileges, even after his release from prison and the end of his parole period, when he is legally presumed to have "paid his debt to society."

Anthropological death may refer to separation from one's usual area of living and working—by self-chosen expatriation, for example, or by legal exile. In a larger sense it refers to the separation of a person from his common or peer group. Knutson (1970) uses the term more narrowly: " 'Anthropological death' refers to one who, like a traitor, has been rejected and cut off from his own community."

Presumptive death, as defined in *Black's Law Dictionary,* is "that which is presumed from proof of a long absence [of a person] unheard from and unexplained. The general rule, as now understood, is that the presumption of life ceases at the expiration of seven years from the time when the person was last known to be living; and after the lapse of that period there is a presumption of death. In most states the subject is regulated by statute."

All these partial deaths represent diminutions and truncations of human talents and creative energies; they are in a way tragic underachievements, slough experiences.

14

A LITERARY EXAMPLE OF PARTIAL DEATH

We have previously discussed various kinds of death, including partial death, and we all know the dictionary definition of death "as the act, or event, or cause, or occasion of dying, or of the end of life, or of cessation." In these terms, by "the death of Herman Melville" one could refer only to a description or a discussion of what occurred to Melville on September 21, 1891; or, by legitimate extension, one might choose to include those events and factors which immediately preceded Melville's physical death on that date, or which were causally related to it. What can one possibly mean by the *deaths* of Herman Melville?*

Melville, in his tortured book *Pierre*,† provided us with an opera-

*This chapter is an edited version of a paper which appeared under the title "The Deaths of Herman Melville" in Howard P. Vincent (ed.), *Melville and Hawthorne in the Berkshires* (1968) and is used with the kind permission of the Kent State University Press.

†The reader is referred to perhaps the most thorough (and, without doubt, the best written) account of *Pierre* and Melville in Henry Murray's Introduction to the Hendricks House edition of *Pierre, Or The Ambiguities* (1949). I cannot say enough in praise of it: that introduction is the best clinical analysis that I know.

tional definition of death as "the last scene of the last act of man's play." But perhaps there is more—"the little lower layer." As we have seen in the last chapter, many deaths-in-life can precede that final irreversible physiological termination and psychological cessation of one's viable existence. Specifically, there are death equivalents, substitutes for death, and especially *partial* deaths, such as deaths of aspects of the self.

In his essay "Dead to the World: Or the Passions of Herman Melville," Murray (1963) asked the following questions:

> Why not some other related condition, such as a temporary or permanent cessation of a part of psychic life—the cessation of affect (feeling almost dead), for example—or the cessation of an orientation of conscious life—the cessation of social life (dead to the outer world) or of spiritual life (dead to the inner world), for example. . . . It might be well to take account of different degrees and of changes of degrees of life—near cessation (as good as dead) or a trend towards cessation (diminution).

The concept of partial death—death of an aspect of the self—is now the pivotal and critical concept for our consideration. Its manifestations are an inner barrenness and aridity, accompanied by withdrawal from his society, grave social refusal, or even where the fires of feeling are still burning, they burn "without glow or warmth," or pleasure-giving purpose. It has to do with repudiation of one's society, of ostracizing people, cutting them dead; it also relates to society's repudiation and ostracism of the person. Thus there are deaths of aspects of the inner self, and deaths of aspects of the outer or social self. Buchanan's failure to find Melville in America—

> I sought everywhere for this Titan who was still living somewhere in New York, but no one seemed to know anything of the one great writer fit to stand shoulder-to-shoulder with Whitman on that continent. . . .

is a poignant example of Melville's not being totally socially alive, a partial social death which lasted for some "forty torpid years."

Melville had an intimate association with the idea of partial death: About it, he might have said: "I am quick to perceive a horror and could still be social with it." It is debatable whether *Moby Dick* is as much about terminal death as it is about partial death, and especially about suicidal equivalents:

> ... whenever it is a damp, drizzly November in my soul ... then I account it high time to get to sea as soon as I can. This is my substitute for pistol and ball. With a philosophical flourish, Cato throws himself upon the sword; I quietly take to the ship.

And again:

> ... to the death-longing eyes of such men, who still have left in them some interior compunctions against suicide, does the all-contributed and all-receptive ocean alluringly spread forth his whole plain of unimaginable, taking terrors, and wonderful, new-life adventures; and from the hearts of infinite Pacifics, the thousand mermaids sing to them—"come hither, broken-hearted; here is another life without the guilt of intermediate death; here are wonders super-natural, without dying for them. Come hither! bury thyself in a life which, to your new equally abhorred and abhorring, landed world, is more oblivious than death. Come hither! put up thy gravestone, too, within the church-yard, and come hither, till we marry thee!"

Before we turn to further explications of these qualitative nuances of death in the human spirit (including Melville's own), I shall, as background material, attempt to present some few quantitative evidences of the extent of Melville's interest in death. I must confess that I present these quantitative data with serious misgivings. To reduce the grand scope of Melville's poetic outpourings to chart and figures can only remind one of Whitman's poem "When I Heard the Learn'd Astronomer":

> When I heard the learn'd astronomer,
> When the proofs, the figures, were ranged in columns before me,
> When I was shown the charts and diagrams, to add, divide, and measure them,
> When I sitting heard the astronomer where he lectured with much applause in the lecture-room,

How soon unaccountable I became tired and sick,
Till rising and gliding out I wander'd off by myself,
In the mystical moist night-air, and from time to time,
Look'd up in perfect silence at the stars.

Eleven of Melville's books—*Typee, Omoo, Mardi, Redburn, White-Jacket, Moby Dick, Pierre, Israel Potter, The Piazza Tales, The Confidence-Man,* and *Billy Budd*—were examined, and a notation was made for each of five types of death material:*

A. Any reference to the death of the chief protagonist, including his wishes about death, his ideation concerning death, threats of death to him (either by himself or by others), or actions which might result in his death.

B. The actual death of specific characters other than the chief protagonists, either as an individual or in a group, such as the entire crew of a ship.

C. Any reference to death relating to other than the chief protagonists: threats of death, actions which might involve the character in destruction, death thoughts, and so on.

D. Death in nature, such as violent storms, and references to the dark, hostile, destructive, death-dealing aspects of nature.

E. Discourse on death, including historic references to naval battles where death occurred and philosophic dissertations on death.

The result of this study was the tabulation of 1,802 references to death (see Table 1). In the accompanying graph the solid line represents the distribution of these references among the eleven works; the dotted line indicates the relative concentration of death talk per page within each work. Of what Melville said about death, he seems to have said most of it—about 30 percent—in *Moby Dick,* followed by *The Piazza Tales, Mardi,* and *Pierre,* with 14, 13, and 12 percent, respectively. The vast majority — 70 percent — of Melville's death thoughts is contained in these four books.

The concentration of death talk per page was computed by comparing the number of death items per book with the number of pages in the Constable edition of that book; then this ratio was compared against the sum of death thoughts in all eleven books. These data

*I am happy to express my appreciation to Miss Jan Kramer, then of the Los Angeles Suicide Prevention Center, for her enormous effort and help in this project.

indicate that there was relatively more death talk per page in *The Piazza Tales* and *Billy Budd* than in any other books, with the next highest concentration in *Moby Dick*.

The distribution of the five death categories among the eleven books is indicated in Table 2. These data show us that most of the items (36 percent) were in the "Discourses on Death" category, the least (7 percent) in the "Death in Nature" category. An interesting exception is *Pierre*, in which 45 percent of the death talk relates to the death of the chief protagonist.

The quality of the references to death varies among the books: *Mardi* is more philosophic, reflecting a fantasied wish for the idealization of society; *Moby Dick* is concerned more with actual death and symbolic substitutes for death as portrayed through the struggle of man with internal and external imperatives; *Pierre* is obsessed with death in the individual psyche; *The Piazza Tales* deals with the death duel between man and nature; and in *Billy Budd* the emphasis is on the interplay of death movements among the chief characters—Billy, Captain Vere, and Claggart.

In brief summary, we found some 1,800 death thoughts distributed among some 4,500 printed pages of Melville's writing, or about one death thought for every two and a half pages. The most death-laden book is *Moby Dick*, next *Mardi, Pierre*, and *The Piazza Tales;* the most death-concentrated books are *The Piazza Tales* (which followed *Moby Dick* by three years) and *Billy Budd* (which followed *Moby Dick* by almost forty years). Death remained in Melville's mind and flowed from his pen from *Mardi* on; that is, from the time that Melville was only thirty years old.

My saying "only thirty years old" would seem to indicate that some break with the past is discernible here. And so it is. The notion of an appropriate time of life, of stages or phases of life in the total human life cycle, has its most famous explication in Shakespeare's seven ages of man. In our own time, the psychoanalyst Erik Erikson has written about eight psycho-social stages in the human life cycle; Jung has discussed the two main ages of man; Gerald Heard has described five ages of man, and Charlotte Bühler has delineated several main stages in the human course of life, to name a few. One main point that all these writers imply is that each major period of life has its special problems, conflicts, crises, ordeals, and mysteries. To give one example, adolescence is a time when the main

TABLE I

DEATH IN MELVILLE

		Among All Books		Within Each Book		
		No. of Items[a]	Percent	No. of Pages[b]	% Thoughts per Page	Percent[c]
Typee	1846	92	5	340	27	6
Omoo	1847	52	3	375	14	3
Mardi	1849	243	13	769	32	7
Redburn	1849	89	5	403	22	5
White-Jacket	1850	109	6	504	22	5
Moby-Dick	1851	538	30	725	74	16
Pierre	1852	218	12	505	41	9
Israel Potter	1855	63	3	225	28	6
Piazza Tales	1856	246	14	271	91	20
Confidence Man	1857	49	3	336	15	3
Billy Budd	1891	103	6	114	90	20
		1802	100%	4567	456	100%

[a] For five death categories: Cessation of chief protagonist, death of anyone, death in nature, discussion of death, and general cessation.

[b] Constable Edition.

[c] Percentage of thoughts per page divided by 4.5 in order to reduce 456 to 100%.

FIGURE I

DEATH IN MELVILLE

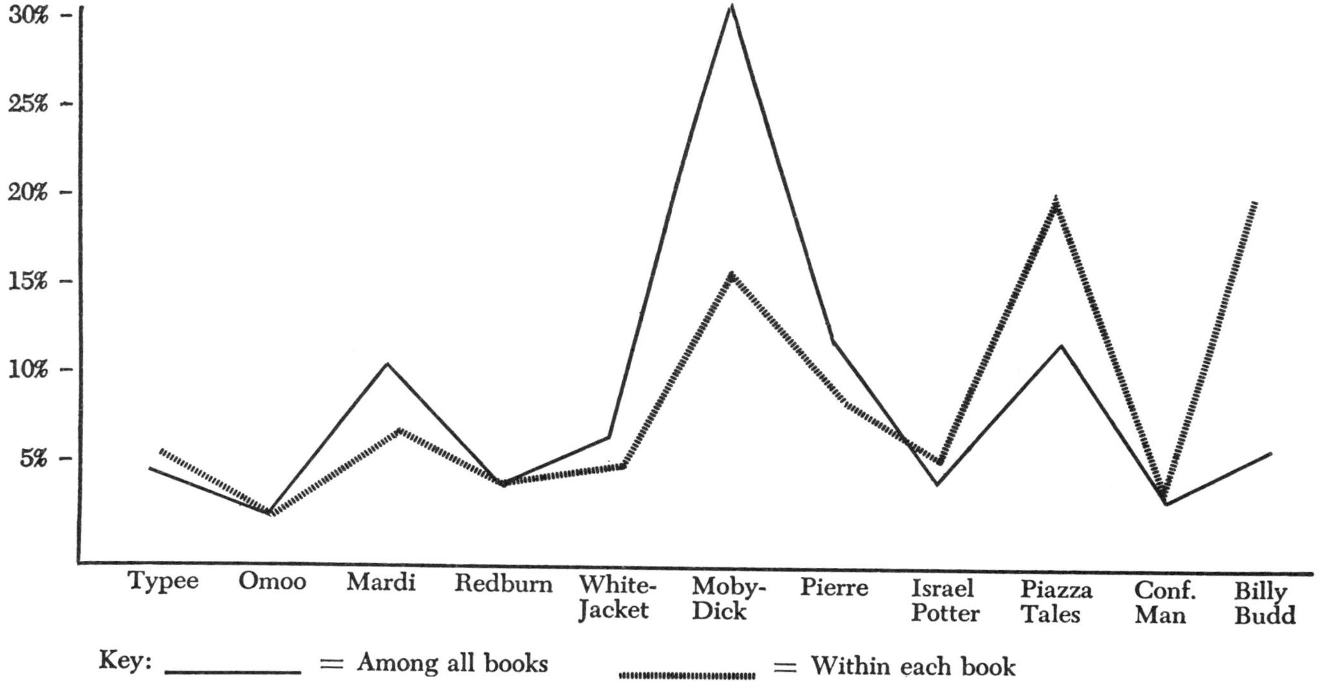

TABLE II

ANALYSIS OF MELVILLE DEATH ITEMS

	Typee	Omoo	Mardi	Redburn	White Jacket	Moby-Dick	Pierre	Israel Potter	Piazza Tales	Conf. Man	Billy Budd
Number of Items, 1802	92	52	243	89	109	538	218	63	246	49	103
Death Categories:											
A. Any reference to death of chief protagonist, including wishes, ideation, threats, etc.	18%	5%	10%	15%	12%	11%	45%	20%	26%	8%	33%
B. Actual death of specific characters other than chief protagonists.	3%	10%	10%	13%	8%	10%	11%	22%	18%	16%	27%
C. Death in nature; death in the environment.	11%	11%	8%	4%	5%	13%	4%	5%	8%	6%	1%
D. Discourses on death; discussion of death; historic references to death.	28%	31%	51%	37%	43%	26%	34%	32%	29%	57%	33%
E. Any reference to death relating to characters other than chief protagonists: wishes, thoughts, threats.	39%	43%	21%	30%	32%	39%	7%	21%	19%	12%	7%
	100%	100%	100%	100%	100%	100%	100%	100%	100%	100%	100%

problem is that of separation from the family and movement into young adulthood, carrying with it the problems of how to find and hold love, the role of romantic love and its relationship to sexuality, how to begin a career, and how to establish and support a family.

But for Melville the time was out of joint: a young man obsessed with an old man's terrors. Had he been able to postpone his confrontation with these enigmatic and unanswerable issues that are more grim even than misery and death—namely, eternal and total cessation, annihilation—until a more proper time in his own life, he might have been a more measured and sanguine soul, but precocious psyche that he was, he encountered these unconquerable phantoms out of phase with his own life, too early.

To answer a soul's need to deal with the issue of one's own annihilation before he is thirty-five is to tackle God's world in off-season, at hurricane time, and then one is faced only with the tortured choice of what nature of wreck will ensue—utter, partial, immobilizing, fatal—not, alas, with the happier alternatives between wreck or safety, between escape or spiritual victory, between destructive or positive affective transactions. Dr. Murray has stated that a "study might be made of the . . . cycle of affective transactions in the works of certain Romantic authors—authors whose credo encourages them to give vent, in apt and telling words, to emotions of all sorts. I need not tell you the name of my first choice among eligible authors."

I have taken what I consider to be a reasonable calculated risk and assumed that it could be no other than Melville whom he had in mind. If this is so, what can we say about Melville's credo?

Melville's credo was to say yes; yes to his own sense of right; yes to his fealties to the depths of his own thought and inner experience; but like any other brave man who feels overcome by vastly superior forces, never to say yes to the commands of others or of fate or of nature to surrender. Melville had several mottoes, for example; *Ego non baptizo te in nomine patris sed in nomine diaboli!*, but his credo seems to have been this: better to be drowned as a daring fool than be hanged as a coward. And this, as we shall see, was just a partial reflection of his great, self-ennobling, courage-building, autonomous sense of pride. He voices this credo over and over. But first, concerning his sense of autonomy:

In *Moby Dick:*

"I'd strike the sun if it insulted me."

"I, myself, am a savage, owing no allegiance but to the King of Cannibals; and ready any moment to rebel against them."

". . . the queenly personality lives in me and feels her royal rights. . . ."

"Here I am proud as Greek God. . . ."

And in a letter to Hawthorne, April 16, 1851:

"The man who, like Russia or the British Empire, declares himself a sovereign nature (in himself) amid the powers of heaven, hell, and earth. He may perish, but so long as he exists, he insists upon treating with all Powers upon an equal basis."

And now, to the core and substance of Melville's credo:

In *Moby Dick*, about Bulkington:

"Better it is to perish in that howling infinite than to be ingloriously dashed upon the lee, even if that were safety."

"Oh man! Admire and model thyself after the whale. Do thou too remain warm among ice. Do thou, too, live in this world without being of it. . . . Retain, O man! in all seasons a temperature of thine own."

In *Pierre:*

"For now am I hate-shod! On these I will skate to my acquittal! No longer do I hold terms with aught. World's bread of life, and world's breath of honour, both are snatched from me; but I defy all world's bread and breath. Here I step out before the drawn-up worlds in widest space and challenge one and all of them to battle."

Jimmy Rose:

"I still must meditate upon his strange example, whereof the marvel is, how after that gay, dashing, nobleman's career, he could be content to crawl through life, and peep about among the marbles and mahoganies for contumelious tea and toast, where once like a very Warwick he had feasted the huzzaing world with Burgundy and venison."

And from his letters:

To Hawthorne, June 1, 1851: "All Fame is patronage. Let me be infamous: there is no patronage in that. . . ."

"To the dogs with the head! I had rather be a fool with a heart, than Jupiter Olympus with his head. . . ."

To Duyckinck, March 3, 1849:
". . . then had I rather be a fool than a wise man.—I love all men who dive. Any fish can swim near the surface, but it takes a great whale to go down stairs five miles or more; & if he don't attain the bottom, why, all the lead in Galena can't fashion the plumet that will. . . ."

Along with his many other ambiguities, Melville, with his great zest for life, was a death-intoxicated man. Thoughts of death appear through his works as a leitmotif, subliminally haunting the reader with their baleful expressions of grandeur and woe. For Melville, it was more than do or die; it was exult or never-have-been. The essence of his credo, from his third book, *Mardi*—and one of my own favorite Melville passages—written when he was twenty-nine, reads as follows:

> Oh, reader, list! I've chartless voyaged. With compass and the lead, we had not found those Mardian Isles. Those who boldly launch, cast off all cables; and turning from the common breeze, that's fair for all, with their own breath, fill their own sails. Hug the shore, naught new is seen; and "Land ho!" at last was sung, when a new world was sought.
>
> That voyager steered his bark through seas, untracked before; ploughed his own path mid jeers; though with a heart that oft was heavy with the thought that he might only be too bold, and grope where land was none.
>
> So I.
>
> And though essaying but a sportive sail, I was driven from my course, by a blast resistless; and ill-provided, young, and bowed to the brunt of things before my prime, still fly before the gale;—hard have I striven to keep stout heart.
>
> And if it harder be, than e'er before, to find new climes, when now our seas have oft been circled by ten thousand prows—much more the glory!

> But this new world here sought, is stranger far than his, who stretched his vans from Palos. It is the world of mind; wherein the wanderer may gaze round, with more of wonder than Balboa's band roving through the golden Aztec glades.
>
> But fiery yearnings their own phantom-future make, and deem it present. So, if after all those fearful, fainting trances, the verdict be, the golden haven was not gained;—yet, in bold quest thereof, better to sink in boundless deeps, than float on vulgar shoals; and give me, ye gods, an utter wreck, if wreck I do.

In any complete recitation of Melville's unusual combination of gifts, pressures, heritage, and inner thrusts, the role of his overweening pride must be given the great emphasis it obviously merits. At the outset, we can, following Nietzsche, distinguish pride from vanity: whereas the former comes mostly from inner feelings, the latter arises from outside reinforcement and feedback, as in the case, for example, of poets. With Melville, it is pride, not vanity, that we mean.

It would appear that Melville's sense of pride stemmed largely from two sources: his family lineage, and his possession of a uniquely creative intellect. His pride in his family is clearly indicated in his books. In *Moby Dick* he writes: "It touches one's sense of honor, particularly if you come of an old established family in the land, the Van Rensselaers, or Randolphs, or Hardicanutes."—explicitly stated in his letters, as, for example, his letter to his mother (May 5, 1870), and to his cousin (August 27, 1876).

As for his pride in himself, as early as 1839, when he was nineteen, Melville wrote in his *Fragments from a Writing Desk* that his mind was "endowed with rare powers"; all through his letters he writes either with the exaggerated modesty of a proud man or the proper self-respect of a proud man. For example, he writes to his father-in-law: "I hope the perusal of this little narrative of mine will offer you some entertainment." He describes himself in his letters as "conceited," "garrulous," filled with "selfishness" and "egotism." He had an enormous appetite for recognition and fame, and eagerly tracked down all the reviews of his works. Melville was obviously deeply wounded by the critics of *Mardi* and *Moby Dick*. After the *New York Literary World's* review of *Moby Dick* he canceled his subscription to that journal.

Now, if one believes in the power of the self-fulfilling prophecy, the role of pride can be a salutary one: it enhances one's self-confidence and increases effectiveness. But the negative side of this coin is that it also makes one vulnerable. The enemy of pride is criticism and rejection. The public, represented by the critics, liked Melville's tobacco and wood-chopping works but railed and hooted when he wanted to say the things most meaningful to him. "What I feel most moved to write, that is banned,–it will not pay. Yet, altogether, write the other way I cannot," he lamented in a letter to Hawthorne in 1851.

And in *The Confidence Man,* a story of aristocratic pride and its concomitant fear of dependency and vulnerability, he writes: "When both glasses were filled, Charlemont took his, and lifting it, added lowly: 'If ever, in days to come, you shall see ruin at hand, and, thinking you understand mankind, shall tremble for your pride; and, partly through love for the one and fear for the other, shall resolve to be beforehand with the world, and save it from a sin by prospectively taking that sin to yourself, then will you do as one I now dream of once did, and like him will you suffer.' "

In a situation where one has been verbally abused and behaviorally constrained, one can generally react in one of two ways. He can protest, but this course will demonstrate how deeply he cares and how grievously he has been hurt. This tack will make him more vulnerable, more pitiable, and often subject to more abuse. It is a course which runs directly counter to the instincts of a proud man. The other course is to be disdainful, ignoring one's critics as though they did not exist, reducing them to impotence by robbing them of their power to hurt. But this maneuver, by its very nature, can be executed only at the price of self-ostracism, and this denial of the psychological self is a type of social death.

In Stedman's Introduction to the 1892 edition of *Typee,* we read a contemporary illustration of Melville's social death:

> Mr. Melville would have been more than mortal if he had been indifferent to his loss of popularity. Yet he seemed contented to preserve an entirely independent attitude, and to trust to the verdict of the future. The smallest amount of activity would have kept him before the public; but his reserve would not permit this. That he had faith in the eventual reinstatement of his reputation cannot be doubted. . . . Our author's tendency

to philosophical discussion is strikingly set forth in a letter from Dr. Titus Munson Coan to the latter's mother, written while a student at Williams College.... The letter reads in part:

"... when I left him he was in full tide of discourse on all things sacred and profane. But he seems to put away the objective side of his life, and to shut himself up in this cold north as a cloistered thinker."

One human emotion not inconsistent with pride is hate or hostility. Melville had more than his fair share of hate, conscious and unconscious: against his parents—his mother and father for separate reasons; against society; and against Christendom—all the forces that he saw as wanting him to acquiesce. Granted his hate was long repressed by his fear of retaliation; but who can gainsay his attacks on Christianity in *Omoo* and on Western society in *Typee*?

We see this combination of pride, fear, and hate in Ahab's address to the candles:

> Oh! thou clear spirit of clear fire, whom on these seas I as a Persian once did worship, till in the sacramental act so burned by thee, that to this hour I bear the scar; I now know thee, thou clear spirit, and I know now that thy right worship is defiance. To neither love nor reverence wilt thou be kind; and e'en for hate thou canst but kill; and all are killed. No fearless fool now fronts thee. I own thy speechless, placeless power; but to the last gasp of my earthquake life will dispute its unconditional, unintegral mastery in me. In the midst of the personified impersonal, a personality stands here. Though but a point at best; whencesoe'er I came; wheresoe'er I go; yet while I earthly live, the queenly personality lives in me, and feels her royal rights. But war is pain, and hate is woe. Come in thy lowest form of love, and I will kneel and kiss thee; but at thy highest, come as mere supernal power; and though thou launchest navies of full-freighted worlds, there's that in here that still remains indifferent. Oh, thou clear spirit, of thy fire thou madest me, and like a true child of fire, I breathe it back to thee.

Nor need we unduly exercise our imagination to divine Melville's attitudes toward his critics:

In *White-Jacket:* "How were these officers to *gain glory*? How but by a distinguished slaughtering of their fellowmen? How were they to be promoted? How but over the buried heads of killed comrades and messmates?"

In *White-Jacket:* "Do you straighten yourself to think that you have committed a murder, when a chance falling stone has often done the same? Is it a proud thing to topple down six feet perpendicular of immortal manhood, though that loft living tower needed perhaps thirty good growing summers to bring it to maturity? Poor savage! And you account it so glorious, do you, to mutilate and destroy what God himself was more than a quarter of a century in building?"

Moby Dick: "There's a most doleful and most mocking funeral! The sea-vultures all in pious mourning, the air-sharks all punctiliously in black or speckled. In life but few of them would have helped the whale, I ween, if peradventure he had needed it; but upon the banquet of his funeral they most piously do pounce. Oh, horrible vulturism of earth! from which not the mightiest whale is free."

In *The Encantadas:* "If some books are deemed most baneful and their sale forbid, how, then with *deadlier* facts, not dreams of doting men? Those whom books will hurt will not be proof against events. Events, not books, should be forbid."

In Moby Dick: "Sharks . . . like hungry dogs 'round a table where red meat is being carved."

And from his letters:

To Lemuel Shaw, April 23, 1848: "I see that *Mardi* has been cut into by the London Atheneum and also burnt by the common hangman in the Boston Post"; and "There is nothing in it, cried the dunce, when he threw down the 47th problem of the 1st book of Euclid—there's nothing in it.—This with the posed critic."

To Evert Duyckinck, December 13, 1850: ". . . I don't know but a book in a man's brain is better off than a book bound in calf—at any rate it is safer from criticism."

To John W. Henry Canoll, January 18, 1886: "For what can one do with the Press? Retaliate? Should it ever publish the rejoinder, they can."

And for pure, undiluted, heartbreaking hate, no other passage in our literature is so concentrated as this one from *Moby Dick*:

All that most maddens and torments; all that stirs up the lees

> of things; all truth with malice in it; all that cracks the sinews and cakes the brain; all the subtle demonisms of life and thought; all evil, to crazy Ahab, were visibly personified, and made practically assailable in Moby Dick. He piled upon the whale's white hump, the sum of all the general rage and hate felt by his whole race from Adam down; and then, as if his chest had been a mortar, he burst his hot heart's shell upon it.

Implicit in all the preceding is a broad hypothesis: that aspects of Melville's personality are to be found distributed (in either a direct or refractory way) among his fictional characters; and further, that the totality of his characters can be made to yield to the totality of Melville's personality. In an article relating to this hypothesis, the Italian critic Eugenio Montale (1960) mentioned, but certainly does not endorse—nor do I—the interpretation that the three principal characters in *Billy Budd* stand as different "narcissistic projections of the three ages of the author." More to the point and better documented is Professor Bezanson's discussion of "The Characters" in his introduction to the Hendricks House edition of *Clarel,* in which he convincingly advances the proposition that several of the main characters were intended to represent different aspects or potentialities of Melville's nature, and specifically that Mortmain embodied Melville's wish for self-annihilation. I quote from Professor Bezanson:

> In the two decades after *Moby Dick,* Melville's descent into self had made him acquainted with an underworld of recalcitrant shades: the sense of defeat, willful isolation, unmanageable moods, fear of death, anxiety over his own physical and mental health. . . . These darker elements of Melville's sensibility are channeled into the striking series of monomaniacs who followed one another so ominously through the poem: . . . From his first appearance Montmain is committed by name ("Death Hand") and symbol (his black skull cap) to *self-annihilation.* The roots of his personal malaise, running subtle and deep, have flowered into political, philosophic, and religious despair. . . . Consumed by psychic fury, driven to *intolerable introversion,* Mortmain has no strength left to hold back his own will to self-destruction.

Every student of Melville is well acquainted with Professor Murray's (1949) explication of Melville's psychological positions over time in relation to his own annihilation. Quoting from Dr. Murray (1949):

> In *Mardi* (1847), Melville's position was: "If I fail to reach my golden haven, may my annihilation be complete; all or nothing."
>
> In *Moby Dick* (1851): "I forsee my annihilation, but against this verdict I shall hurl my everlasting protest."
>
> In *Pierre* (1852): "I am confronted by annihilation, but I cannot make up my mind to it."
>
> In 1856, in London, as recorded by Hawthorne: "I have pretty much made up my mind to be annihilated."
>
> And in *Billy Budd* (1891): "I accept my annihilation."

After Melville's early success with *Typee* and *Omoo*, when he then dared, in *Mardi*, to explore the infinite perspectives of his private universe and discovered that he was met with a barrage of annihilating criticism, the issue for him was forever joined. Thereafter, his most vital dialogue of the mind concerned itself with whether or not he could fully live, express himself openly, be read widely, be adjudged fairly. The choice was clear: he would need to eschew critical support, retreat from public view, write to his own soul's need—all *partial* deaths of his talents, his energies, his interplay with his own time.

Out of Melville's great pride, his bravura inner life style, his imperious reaction to criticism, his withdrawal from social life—out of all these grew his concern with death and annihilation and his enormous investment in his postself. In this sense, Melville wrote not so much for his own time as he did for any appropriate (that is, any appreciating) time to follow—for what Leyda has called a "posthumous celestial glory." Melville was partially dead during much of his own life, but he more than compensated for that lugubrious limitation by writing in such a way that he could realistically expect to live on in the literature of the future. "On life and death this old man walked."

15

MEGADEATH: CHILDREN OF THE NUCLEAR FAMILY

Death is in the air. Fear of personal and mass annihilation pervades thought and emotion, touches the conscience, and influences behavior. There is persuasive evidence that this thanatological aura of our time can be traced to an evil that is "visibly personified and made practically assailable" in the omnipresent threat of nuclear destruction. The Bomb has at last made it impossible for us to deny or ignore man's irrationality and capacity for evil.

It would be surprising if the existence of the Bomb were not deeply inimical to the minds of men. The very threat posed by the historical precedent of Hiroshima and Nagasaki is enough to distort many lives. The psychological fallout from yet unexploded bombs has created a chronic low-grade psychic infection throughout the world. And every day the fear of sudden mass death is exacerbated by new tests of "improved" models in the arms race.

Death appears to be discussed more openly now than at any time since the days of the Black Death in the fourteenth century, when a giant fear cast its shadow over men's lives—but with this difference: the plague was a "natural" catastrophe and thought of as the "will of God"; the Bomb is the will of man. Today, man is the enemy. It is man who threatens his own destruction. We are afraid of our own power for evil and our seeming inability to control it in others.

We talk about the nuclear death with which we are threatened, we are impelled to talk about it. Yet at the same time, as we have seen, the subject of death is taboo. How are we to account for this? And how do young people feel about these issues? Some commentators have said that the threat of sudden mass death matters little to youth; others have said that they cannot escape this specter. My own belief, one reinforced by my students' responses to the questions I put to them, is that much of this new dual attitude toward death is either directly or indirectly related to the Bomb.

More than a decade ago Herman Kahn, in his book *On Thermonuclear War* (1960), presented some grisly arithmetic on the effects of nuclear war in terms of the prospects of rebuilding. (It was Kahn who coined the word "megadeath.") He reasoned that with so many millions killed in this country we could rebuild and reach a gross national product of so many dollars in so many years under certain conditions. A number of other books (Feifel, ed., 1959; Choron 1963; Hinton 1967; Fulton, ed., 1965; Lifton 1967; Vonnegut 1969) and movies (*On the Beach, Dr. Strangelove*) have emphasized the human implications and consequences of massive destruction by bombing. These books and films have had a prodigious impact on our current attitudes toward death. Keniston, in *Young Radicals* (1968), put it this way:

> In the lives of young radicals and their generation, the threat of outer violence has been not only a backdrop, but a constant fact of life. It is reflected not only in childhood terrors of the Bomb, but in the routine experience of air-raid drills in school, to constant exposure to discussion of fallout shelters, preventive warfare, ballistic missiles, and anti-missile defenses, and sometimes in a compulsive fascination with the technology of destruction. The Bomb and what it symbolizes has set the tone for this generation, even for the majority who made a semi-deliberate point of trying not to think about it. There are relatively few young Americans who, upon hearing a distant explosion, seeing a bright flash of light, or hearing a faraway sound of jets overhead at night, have not wondered for a brief instant whether this might not be "It."

One cannot fail to include this overhanging threat of violent mass death as one of the vitally important items in the total array of forces that, consciously or unconsciously, have shaped our current views of life and death. Moreover, the intellectual and philosophical climate of this century has undoubtedly served both to stimulate and to reflect the dysphoric spirit of our time—the century of World Wars I and II, the German death camps, Coventry, Dresden, Hiroshima. Whether for good or for ill, some substantial portion of our current concerns with death must be attributed to Kierkegaard, Jaspers, Heidegger, Marcel, Sartre, and Camus, those existential philosophers for whom the topic of death has often occupied a central place—what Jaspers described as the "awareness of the fragility of being." For Heidegger, being-in-the-world was, in its essence, a being-toward-death. Sartre wrote that "death became for us the habitual object of our concern." And Camus began his *Myth of Sisyphus* with the declaration that *the* central problem for all philosophy was no less than that of death by self-destruction: "Judging whether life is or is not worth living amounts to answering the fundamental question of philosophy."

Sartre's words encompass the central point: death is now the habitual topic of our concern. If death is not the focus of our conscious anxieties, it is, at the hauntingly least, the omnipresent threat that mobilizes our deepest fears.

The step from thoughts of death to thoughts of war is a short one. War is heinous and wasteful. We all know that. But is the war in Indochina more shameful and immoral than others we have waged? What is most painfully shameful is that is is only one more variation on a theme. Our atrocities in Vietnam are bloody repetitions of similar ferocities against Mexicans in 1846 and 1847, against Indians a hundred years ago, and against Filipinos at the turn of the century: entering territories not our own, burning villages to get at native insurrectionists, slaughtering old men, women, and children for revenge or sport, changing and perverting the indigenous culture. Familiar too are the postwar dysphoria and ennui: recall the jaded expatriates after World War I, with their disillusionment, fear, anger, despair, and fitful productivity.

Among the current young generation, many perceive the threat of mass death as very real; they cannot be assured of a future. Having lost this hope, they lack that essential future-tied perspective, that optimism which spurs people to productivity, to goal-directed actions, and, in the end, to life. It is what Alfred Kazin speaks of as the "obstinate hopes of mankind." One needs these hopes to keep going. To lose hope is to become frightened and embittered, then angry and desperate.

Nowadays the prospect of sudden death—by nuclear devices, counterinsurgency, police action, and war—is everywhere. Such notions as megadeath, overkill, and global annihilation have created an urgency to grapple with the topic of death. With nuclear bombs, man now has the capacity to literally erase both the past and the future in one flash of lethal heat. The potential enormity of the superbombs is not only that they are capable of killing practically everyone who is alive and thus jeopardizing the future, but that in doing so they would also "murder" history: all previous ages would be simultaneously erased.

Elliott, in his extraordinary book, *Twentieth Century Book of the Dead* (1972), states: "Since the atom bomb and its successors in death technology, the new possibility of total death for the species has become a reality." He continues to explicate this haunting notion of total death:

> Violence in the twentieth century has produced the new phenomenon of total death. As an *idea*, total death has existed—in mental pictures of the day of judgment, doomsday, the end of the world—at least since the formulation of the great religions. As a *reality* attainable by human means, the science of which is a permanent unalterable part of knowledge, it originates in the notorious half-century from which we are just emerging. *Can* we emerge from the nightmare of reality and vision created in that period? We cannot create a retrospective order for the chaos of the actual events. Can we escape from the chaos of the idea that is left to us? Total death could mean the obliteration of particular cities or countries or religions; it could mean the collapse of world civilization or the death of the species; or it could mean the total death of the mind within a variety of physical parameters. Total death might be brought about by a wide range of means: by the carefully considered destruction of selected millions; by the direct and secondary effects of pollution or overcrowding; by a death-breeding mixture of every kind of human motivation acting on machineries and systems which are beyond the control of living creatures. Total death has a timespan overwhelming the convenient human notion of time. It can 'happen' in an instant, in a few

days; it can have the monthly, yearly rhythms of traditional warfare or it could create a chronic long-term disruption of the seasons of nature and the years of human life. Its possibility is tomorrow, or in the next two hundred years, or at any 'time' in the future. Total death is a hard, scientific and immediate reality at the same time as being a speculative idea in search of a philosophy. No existing mental structures, of science, philosophy or religion, are adequate to contain it.

Robert Lifton, in the first chapter of his remarkable and searing book about the survivors of Hiroshima, *Death in Life* (1967), says:

> With Hiroshima (and her neglected historical sister, Nagasaki) something more is involved: a dimension of totality, a sense of ultimate annihilation—of cities, nations, the world. The feeling may be vague, but it is of the greatest psychological importance. What I am suggesting is that our perceptions of Hiroshima are the beginnings of new dimensions of thought about death and life.

Just as "the nuclear weapons left a powerful imprint upon the Japanese which continues to be transmitted, historically and psychologically, through the generations," those same detonations created in some of my college students (born about that time) a preprint that influenced their lives through a fallout of apprehension and fear. An interesting question is whether many of the current generation might not be termed psychological *Hibakusha* ("explosion-affected persons"). At the least, they are Atomic Bomb Age children. Their comments, then, are not to be lightly dismissed as adolescent nonsense. Here are some of them, contributed by students in my death classes at Harvard and UCLA.

> I think my whole generation has grown up and been tremendously influenced by the imminent threat of death. Perhaps this has promoted somewhat of an attitude of abandon. It certainly contributed to the aura of violence and so has the war. I think that the cloud of destruction has made many people feel the urgency of the need for reform in our own society. Things move more quickly toward change because life could be ended tomorrow. [Female, nineteen]

> When I was quite young I was afraid to stay in New York because I knew that in the event of a nuclear war it would be destroyed. [Male, eighteen]

> I'm sure a week doesn't go by when I don't seriously consider the possibility of nuclear catastrophe. You can't read the paper or *Time* magazine without getting it drummed into your head. I used as a child to be struck dumb with the thought of being killed by an atomic bomb, but as I grew I realized that if I remained in cities like New York, Boston, or London I would hardly have the time to realize I was going to be dead before I would be. [Male, twenty-one]

> The possibility of massive destruction by nuclear devices I am sure has changed everyone's outlook. As for me, I find that such an event is generally impossible to visualize, a thing so incredibly monstrous in every way that it is hard even to think about it. When I do think about it, it often sends me into very deep depression, a hopeless feeling of "What's the use?"—one I suppose which might even lead to suicide. However, this specter has also given me a different perspective on life. In order to live with this thing hanging over you, you have to live as though it were not there, but the fact that it is there gives life a much more immediate aspect to it. It is much more important to live a life which is meaningful and which is good. Also, since there is nothing that the individual can do to stop a nuclear destruction of the world, it kind of makes one accept death as something which will happen, and which should not be taken too seriously. Live life for today and not worry too much about tomorrow. [Male, nineteen]

At this point, in order to orient ourselves temporally, we need to interpose some chronology, remembering that the atomic bombs were dropped on Hiroshima and Nagasaki in August 1945 and the era of bomb shelters and air-raid drills was around 1950 to 1958. These students were writing during the 1969-1972 academic years.

> When I was about fifteen I would dream about the world ending or having to run to our fallout shelter, but this was a time when everyone was talking about building family shelters.

It was only with recent ABM discussion that the threat of nuclear destruction has become real to me once more. [Female, twenty-one]

As a child of eight or nine, I sometimes had bad dreams of bombing raids and running to shelters, something you would not expect of a child who had never had any sort of war experience. But at that age I read newspapers and I listened to newsbroadcasts, so I realized with a fair amount of sophistication the transitory nature of any security. [Male, twenty-one]

I remember being terrified one night when the newly installed air raid siren wailed the take-cover signal. Usually it only sounded the all-clear and that was at 10 A.M. Saturday. This was the most terrifying thing I remember. Nobody could convince me it was a practice. I felt I would have known. That started me thinking about the relative safety of our small town; "Maybe they can't see us from the air?" [Male, twenty-two]

A theoretical issue: What effect does the temper of the times play on attitudes toward death? We all know that the content and degree of religious fervor, beliefs, and superstitions, together with generally accepted ideas, all directly influence basic attitudes toward death at any given time. For contemporary youth, the lifelong spirit of the time has been the spirit of war.

I fear war. I despise, for the most part, those who fight. I see absolutely no justification for starting or continuing a war. Nuclear war scares the shit out of me. I don't think it makes me, personally, more afraid of death, but it does make me worry about the death of a massive number of people. I think the greatest influence the nuclear bomb has had on me is to shock me. I can't believe that man can be so stupid as to want to kill himself three times over. [Female, eighteen]

I feel very strongly that war is totally wrong, under any circumstances, something that I might not believe if the date were 1939. [Male, eighteen]

The bomb has made me conscientiously opposed to all forms of war. It has put a value on the now, the me, the feelings, not the job, the money, the nice life. [Male, twenty]

The winds of war—seemingly endless and pointless war—and the gales of disappointment and adversity have blown through their generation. One more (for whom the forecastle was his Harvard): "I had learned to think much and bitterly before my time. Cold, bitter cold as December, and bleak as its blasts, seemed the world to me then; there is no misanthrope like a boy disappointed and such was I, with the warm soul of me flogged out by adversity." So said Melville's Redburn as he set out on his life voyage over a hundred years ago.

Traditionally in the face of "ordinary death"—hospital quiet and death-room dignified—man becomes compliant, law-abiding (obeying everything from nurses' strictures to the laws of nature), and "generatic"—that is, thinking of his benign influence on generations to follow. But under the psychological goad of this new heinous threat of global annihilation, man may turn on his tormentors and pounce on hapless scapegoats. Fear breeds paranoia and paranoia breeds violence. In *Narcissus and Goldmund,* Hesse vividly describes how during the Black Plague the populace turned savagely on the Jews in their midst, burning and torturing them in fear-crazed desperation.

> The whole region, the whole vast land lay under a cloud of death, under a veil of horror, fear, and darkening of the soul. And, worst of all, everybody looked for a scapegoat for his unbearable misery. Whoever was suspected of these horrors was lost, unless he was warned and able to flee: either the law or the mob condemned him to death. The rich blamed the poor, or vice versa; both blamed the Jews, or the French, or the doctors. In one town, Goldmund watched with grim heart while the entire ghetto was burned, house after house, with the howling mob standing around, driving screaming fugitives back into the fire with swords and clubs. In the sanity of fear and bitterness, innocent people were murdered, burned and tortured everywhere. Goldmund watched it all with rage and revulsion. The world seemed destroyed and poisoned; there seemed to be no joy, no more innocence, no more love on earth.

At such times the gloves are off, the usual rules abrogated. Incivility occurs easily, then open rebellion and lethal hostilities. This is what is happening today in our cities, our schools, our prisons. In order to de-escalate this dangerous rage, our primary need is to assuage atomic fear and undo the great psychic harm created by our long-term sanction of this evil. Only then can we hope to live peaceably within our borders. Measures short of that—changing the draft to a lottery, bringing some of the boys home from Vietnam (while flying others to Laos), ineffectually desegregating some schools, inadequately building some public housing—are, from youth's point of view, a sardonic Band-aid on a hemorrhaging wound.

We live in a fulcrum time. Youth sees the current world scene as a life-and-death seesaw, with life teetering in the balance. But this is not the view of youth alone. In a recent article John Platt (1969) of the University of Michigan spells out in a dramatic and terrifying manner the root problems of our age, and, in the absence of their solution—a subject on which he is not resoundingly optimistic—the 50-50 chance of our surviving until 1990. "As long as we continue to have no adequate stabilizing peacekeeping structures for the world, we continue to live under the daily threat not only of local wars but of nuclear escalation with overkill and megatonnage enough to destroy all life on earth." For both the United States and the world as a whole, "the one crisis that must be ranked at the top in total danger and imminence is, of course, the danger of large-scale or total annihilation by nuclear escalation or by radiological-chemical-biological warfare."

We are at such a fulcrum time in the psychological history of man, "vibrating in mid-deep," debating (nuclear destruction, overpopulation, environmental pollution) whether to sink into "the speechless profound of the sea" or to bound up into the blessed air, whether to respond to the imperious call of life or to succumb to the constant invitation that death seductively puts forth. In Platt's survival chart, in the grid where "total annihilation" intersects "20 years from now," he has put a black German iron cross, and in his text this comment: "The peace-keeping stabilization problem will either be solved by that time or we will probably be dead."

Death was put in the air by the older generation. It is they who have created the shameful state of affairs for which they are accounted morally responsible. It is the older generation that threatens the life of today's youth. Authority's mistakes are youth's burdens. The unfairness of it all makes youth angry and disrespectful. Civility to authority goes out the window because approbation is not sought. And with the collapse of civility, all degrees of hostility are then psychologically possible. It depends only on the provocation, the devised occasion.

If the young people do not identify with their elders' causes, where then are their cathexes? Their psychological investments are in their own acts, their fight, the youth crusade. What keeps it going is a very basic emotion: excitement. Excitement gives a keen edge to their sense of vitality, bravado, and exploration. This generally heightened derring-do leads to heightened risk-taking behavior; with motorcycles, confrontations with police, suicide attempts, promiscuity, drugs. And so for current youth the specter of death now hovers at two doors: the first unbeckoned with its nuclear armload; but the second, teased and flirted with. The paradox is that in youth's concern to forget that death is in the air, some have put death in their actions.

> I don't think that the possibility of nuclear destruction has added to my own death fears, although they are a function of the alienation, callousness, competitiveness of twentieth-century society. As far as an idea of an end to human history is concerned, occasionally, when I am in a historic, time-flow vastness of-the-human-struggle frame of mind, the incredible disproportion, unfairness, cosmic quality of such an end boggles my mind. [Male, twenty-one]

> The thought of nuclear war appalls and saddens me. That the possibility for such large-scale destruction of the gift of life exists and is allowed to exist and, perhaps, must exist, fills me with awe and a chilling sense of dread. Nuclear war would mean more than just my personal death, which is really not so important—it would mean the death of the whole human race and everything we have tried to accomplish. If we really were created by God—and I do feel He had a hand in the creation of each of us—then we would certainly turn out to be a poor experiment if we destroyed ourselves. [Male, nineteen]

> The constant threat of nuclear war, the arms race, and the continual reminder of this threat on television makes me acutely aware that death could happen at any time. It magnifies the

instability that life has in the face of inevitable death. Sometimes I feel that this whole process of rapidly advancing technology, by far surpassing any sort of philosophy, has begun a chain of events that will lead to the destruction of mankind. It is almost as if this were the work of natural selection again, where man has gone so far in trying to manipulate nature that he will destroy himself. In this way, death can be seen as the natural process of maintaining an equilibrium of natural forces. It is very hard in light of all the technology that man has, his small amount of knowledge of the universe, and his tendency to desire to destroy other men, to have any belief in the benevolent and just God, who ordains that "the meek shall inherit the earth." Because of this, the comforts of religion, and afterlife, and an interceding God lose much of their credibility and make the prospect of death that much worse. [Female, eighteen]

For this generation there is a heightened sense of the uncertainty of the future, of the survival of the world itself. No other generation in memory has grown up with this particular global tenuousness. Young people today do not worry so much about the nature of tomorrow (as, for example, those who lived through the Depression had to do); they rather worry about whether or not there will be a tomorrow. For the first time in six centuries (since the great European plagues) a generation has been born and raised in a thanatological context, concerned with the imminent possibility of the death of the person, the death of humanity, the death of the universe, and, by necessary extension, the death of God. In 1962 Gunther Anders wrote:

If there is anything that modern man regards as infinite, it is no longer God; nor is it nature, let alone morality or culture; it is his own power. *Creatio ex nihilo*, which was once the mark of omipotence, has been supplanted by its opposite, *potestas annihilationis* or *reductio ad nihil;* and this power to destroy, to reduce to nothingness, lies in our own hands. The Promethean dream of omnipotence has at long last come true, though in an unexpected form. Since we are in a position to inflict absolute destruction on each other, we have apocalyptic powers. It is we who are infinite. There is a new general threat hanging

> over us. Its implications have become even more sinister: for what is exterminable today is not "merely" all men, but mankind as a whole. This change inaugurates a new historical epoch. Accordingly, all history can be divided into three chapters with the following captions: (1) All men are mortal, (2) All men are exterminable, and (3) Mankind as a whole is exterminable.

And one might add a fourth chapter, captioned "God is exterminable."

What are some of the effects of this lugubrious death consciousness on students? The effects seem to be varied and even contradictory: a heightened emphasis on the present (the "now generation"); hopelessness and resignation, a waiting for the end; indignation over one's impotence to control the conditions that threaten mass death; even an urge to get it over with. There is taboo and permissiveness; repression and heightened consciousness; romanticization of death and sardonic realism.

A sizable minority of the students in any class on death—about 25 percent of the females and 35 percent of the males—indicated that the Bomb had little or no influence on their attitudes toward death. A few simply discounted any possibility of a future atomic war:

> I can't believe it would really happen so it hasn't really had any influence. [Female, twenty]

> The possibility of massive destruction has done nothing to my attitudes about death because I simply don't accept the possibility. [Male, twenty-one]

A handful of students expressed the curious attitude that because one has to die sometime, the time and manner of death, whether by atomic bomb or otherwise, made little difference:

> The possibility of massive human destruction by nuclear war has not influenced my present attitudes toward death. If destruction comes, I'll accept it. Everyone must die sometime. The method is not at all relevant or important. [Male, twenty-one.]

> If it has had any effect, I'm not aware of it. I never give a

> thought to the fact (and I do recognize it as a fact) that nuclear war could kill me at any time. My death will be the same experience whether it comes when I'm twenty or seventy. [Male, eighteen]

> The nuclear bomb is merely an extension of my feeling toward the inevitability of death. There is no use worrying about something over which one has no control. If it happens, it happens. [Male, eighteen]

By a turn of logic, a few students discounted the influence of the Bomb on their thoughts of death by virtue of the fact that they would not, in any event, survive its use:

> I do not feel that nuclear weapons will be utilized and that if I am wrong no one will be around to say "I told you so." [Male, twenty-two]

> The idea of massive human destruction is usually ignored by me. I feel there is no reason for worrying about this, especially since if ever there is a nuclear war, very little chance will be left for survival. [Male, twenty]

Reactions of relative indifference among some students centered around their incapacity to influence the decisions relating to use of the Bomb:

> The awesome power of the bomb is held by the elite; I am not the elite, so I can't influence the possibility of destruction. That bothers me, but only when I think about it. [Male, nineteen]

For some, the threat of nuclear war moved them in the direction of social reform, civil protest, or political participation:

> The possibility of massive human destruction has been the single greatest influence on my present attitude towards death. Mass destruction eliminates the possibility of my quitting society. I am forced to participate in the controlling body of society in order to get control of my own life. The knowledge

that I and the world can be eliminated at any given moment gives a sense of urgency to everything I am involved in. [Male, twenty]

I think my whole generation has grown up and been tremendously influenced by the imminent threat of death. Perhaps this has promoted somewhat of an attitude of abandon. It has certainly contributed to the aura of violence, and so has the war. I think that the cloud of destruction has made many people feel the urgency of the need for reform in our society. Things move more quickly toward change because life could be ended tomorrow. [Female, nineteen]

As with other issues in relation to war and the Bomb, we are dealing with members of a totally committed generation, and never more committed than when they espouse pacifism. The "dirty little war," the draft, the military's devaluation of man—all have had their effects on the hearts and minds of youth. On these issues the older generation seems to youth to be not only two-faced and incompetent, but possessed of a power madness capable of destroying us all.

I don't think about this except when I'm taking part in a political demonstration. It is then that I understand that my country has no real desire to protect me. Perhaps just the understanding that the purpose of those weapons is to destroy people who may very well think the way I do, and to destroy me. [Female, eighteen]

A few students wrote of the role of nuclear death as part of man's total evolutionary history and of the effect of this notion on their own beliefs:

Death seems a great deal more imminent. There are times when I wonder if I shall have a middle age or if it is fair to have children. However, I worry just as much about the massive human destruction of poverty and think it more dangerous. If the whole world is destroyed, then who will be left to care anyway? But as long as we are conscious and watching, how can we be so callous and selfish? Also I have sometimes thought that massive human destruction by nuclear war may be the social means by which evolution is working its course to re-establish a balance which we have thrown askew. [Female, twenty-two]

Not unexpectedly, a large number of students spoke of the threat of nuclear war as influencing their attitudes toward death itself:

> The atomic bomb has secularized death and made it meaningless. [Male, eighteen]

> I consider this mass destruction very likely and hence ugly torturous death is always present in the background. [Male, twenty]

> It has made me more aware of the possibility of death and brought me to accept it in stride. Unfortunately, death is almost too close and too acceptable to me now. Death is a part of my life. Death used to seem too remote and too isolated. Now it is ever present and I contend with it much more frequently. [Male, twenty]

A substantial number of students wrote of how the threat of nuclear death had influenced their attitudes toward *life*. Two patterns emerged, one pointing toward a decreased value of life and the other toward a sense of urgency and deep commitment. Here are some responses that speak of a lessening of the value of life itself:

> I remain optimistic enough to believe that humanity will not be wiped out by nuclear war. But I resent and am fearful of the implications of such possibilities, i.e., a decreasing value placed on the miracle of life itself. [Female, eighteen]

> I suppose they have somewhat depersonalized and dehumanized the concept of death. Life no longer seems such a sacred, personal, individual thing when people by the millions can be exterminated. An extension of this—since life has lost much of its individuality because of this threat, a system such as the ABM has little or no place in our world. What good is life if half of our population is destroyed? Who cares whether you save a few million? Who cares if you exterminate the enemy? Life for the survivors would not be worth living. [Male, twenty]

Several students wrote of a sense of hopelessness and futility induced by the threat of nuclear destruction. They described life as having been made "fragile" and "tentative"; they described themselves as "fatalistic" and even "nihilistic."

> I am less concerned about planning for the future—life insurance, savings, etc.—and more interested in living for now, horrified at the thought of nuclear war but accepting the very real possibility that it may happen. [Male, twenty-one]

> It has made me more bitter about the futility and absurdity of life, when it can be so easily extinguished by one incendiarist, by one pyromaniac. [Female, twenty-one]

Some of those who spoke of their heightened sense of urgency did so with a touch of either ennui or hedonism that came very close to a sense of hopelessness:

> Only makes me more convinced of the advisability of living life a moment at a time, irrespective of the last moment or the moment to come, best letting each moment stand full and complete in and of itself. [Female, eighteen]

> Sometimes I think the possibility of instant holocaust has changed my attitude to life more than my attitude to death. Life must be lived fully now; death should be a catalyst, hanging overhead—a stimulus to make one value life more. [Male, nineteen]

> It has influenced my attitude toward life. This death being so arbitrary and impersonal, the only way I see of coping with its threat is to live an extremely personal, individualistic life, so that when death comes it will be meaningless next to the creativity and enjoyment which has filled my life. Nuclear war has made me feel that life in a movement or death for a cause is nonsense, prostituting one's basic humanity to live as an individual, since nuclear destruction is the epitome of what groups can do. So I think I should like to live as though every day were my last (though this is a difficult achievement, to

make it good enough). And I think that the advent of the possibility of nuclear destruction has affected many people in my generation this way, making our life patterns of expectation and self-denial far different from those of our parents. [Female, twenty]

Here it seems most appropriate to quote Toynbee (1969), speaking out at the age of eighty on his deep concern over the future of the world:

> The greatest change for good is in the increase of social justice in Western countries. There has been a change for ill too—the worldwide exasperation, boiling over into murderous violence, and the application of science and technology to manufacture fiendish and devastating weapons. I have eleven grandchildren and one great-grandchild. They stand, in my feelings, for all the 77,000,000 generations waiting to be born during the next 2,000 million years for which this planet will still be habitable. Question: Are my generation and my children's generation going to let them be born? Or are these two generations going to liquidate the human race?

Professor George Wald (1969) shares Toynbee's concerns and believes that his students do so also:

> I am a teacher at Harvard. I have a class of about 360 students, men and women, most of them freshmen and sophomores. Over these past few years I have felt increasingly that something is terribly wrong. I think I know what is bothering the students: What we are up against is a generation that is by no means sure that it has a future. That is the problem. Unless we can be surer than we are now that this generation has a future, nothing else matters. It is not good enough to give it tender, loving care, to supply it with breakfast foods, to buy it expensive education. These things don't mean a thing unless this generation has a future. And we're not sure that it does.

I shall let an eighteen-year-old Radcliffe undergraduate sum up the feelings of her generation:

> I can't say that I think this or do such and such because we may all die tomorrow, but I'm convinced that in all our minds there's an underlying nagging fear that the world could blow up, a fear which today is more silent than in the bomb-shelter era of a few years ago but perhaps even more strongly felt and exhibited daily in minute, imperceptible ways. Our parents say, "You don't know what it's like to suffer. You didn't live through a depression and a world war." But all of us have had to grow up with the constant knowledge that we could *all* die at any moment, and it's not a specific concrete problem. We can't fight the possibility of destruction with the New Deal or with anti-German artillery. And so, what is most frightening of all, we learn to live with it, we learn to accept it.
>
> What's it like to live in a world which, although besieged by war and poverty, will still assuredly be there tomorrow? We can ask our parents that question. No one can ask us. And so I, and I think all of us, look at death a little differently. It becomes a universal concept, not the death of an individual, of me or my family, but the death of an entire race, an entire world, of all existence. The word death takes on an extra dimension of horror and sadness and unmeasurable frustration and complete and eternal impotence because we'll never be able to reduce it to its former individual meaning.

In 1930 Professor Murray, writing about the Harvard Psychological Clinic, described the death of its founder, Dr. Morton Prince, in this telling way: "And at the end he did not come to death as a shock of corn to his season, but death came to him as a stroke to an old warrior still flushed with anticipations of even greater conquests" (Murray 1957). That phrase "to his season" tells us that there is a time, a "right time," for everything in life. Additionally, it implies that for those special concerns that persist throughout a lifetime (security, dependency, sexuality—and death) there are special adaptive patterns at each stage of life. Erik Erikson has written so lucidly and persuasively about these psychosocial stages of human development that they are now part of our common psychological beliefs.

In *Essays in Self-Destruction* (1967) I called those out-of-phase crises *extratemporal* crises—those that occur without regard to one's time in life—distinguishing them from both intratemporal crises, those that occur "appropriately" within a phase of life, and intertemporal crises, those that occur in the interstice or turning point between two major phases of life. Many of our present youth are, in these terms, out of season, in extratemporal shock.

There is a propitious or appropriate time for the major traumas and tragedies of life to occur—if they must. When they occur too early they are, by definition, "out of phase," and it is forever more difficult for that individual to integrate them into a healthy personality. Melville, in *Redburn*, has said it perfectly:

> Talk not of the bitterness of middle-age and after life; a boy can feel all that, and much more, when upon his young soul the mildew has fallen; and the fruit, which with others is only blasted after ripeness, with him is nipped in the first blossom and bud. And never again can such blights be made good; they strike in too deep and leave such a scar that the air of paradise might not erase it. And it is a cruel and hard thing thus in early youth to taste beforehand the pangs which should be reserved for the stout time of manhood, when the gristle has become bone, and we can stand up and fight out our lives, as a thing tried before and foreseen; for then we are veterans used to sieges and battles, and not green recruits, recoiling at the first shock of the encounter.

Experiences that come too early put one out of kilter with one's own years. These individuals suffer from "information overload" or what might be called "stimulus inundation." The age of three is too young to be orphaned; seven is too young to be seduced; nine is too young to be made aware of giant city-killing bombs, and to be the first generation to have to live with this knowledge can be an especially heavy psychological burden.

The youth of today grew up thinking about death, an orientation discordant with their time of life. Some were in tune with their years; some, either psychologically precocious or retarded, were not. One can, of course, think of death at any time in his life from early childhood on—and young people typically think romantically about death—but, in general, the depth with which death is contemplated reflects the maturity of the thinker. Individuals who are not yet fully born cannot fully die; and the coming of age differs for different people. Herman Melville in a letter to Hawthorne dated his very existence only from the time he was twenty-five: "My development has been all within a few years past. I am like one of those seeds taken out of the Egyptian Pyramids. Until I was twenty-five, I had no development at all. From my twenty-fifth year I date my life." In the first volume of his autobiography, Bertrand Russell (1967) tells us that he first awoke to the most important aspects of his own feelings only when he was in his forties, and then as a result of a world war and a cataclysmic love affair.

Reflections on death by some young people are, for them, premature, because they come before the individual's full seasoning as a youth and are thus precociously out of phase. But most young people are capable of contemplating death. At any stage of life, one is able to contemplate death meaningfully only if he is truly alive in those ways that are appropriate for his chronological age, and only if he has appropriate awareness of his own potentialities and his own finiteness. These insights are prerequisites to any meaningful comprehension and incorporation of the concept of death. For many of us their attainment takes much time—often more time than there is in a life. Youth, too, must be given this time. That is why the cardinal responsibility of the mature to their young is to guarantee them a reasonable chance to survive.

AFTERWORD

The hardy reader who has come this far and who has (either willingly or reluctantly) agreed with most, many, or some of the ideas presented in these pages will, I believe, have a view of death which can enliven his outlook on life. A young Berkeley poet, dying of leukemia at age thirty-one said, from his view, that we fear not death, but the incompleteness of our lives. We need a view which includes an increased appreciation of the role of death in a person's life; an intense respect for the toll of death but, with it, a decreased neurotic obsession with the threat of death. What a horror life would be if ultimately (like for Tithonus) there could be no death—and yet, for many, death represents the ultimate horror of life. Death is an extremely punishing taskmaster, but only if one insists on attending a morbid school. Matriculate at a life-oriented academy, then death can be taken as a necessary and even fascinating part of the total curriculum, an inevitable ending, what some might even choose to call a commencement.

Several of my close friends on whom I pressed the manuscript of this book reported that they found it "heavy stuff," "a real mind trip," that it had created some distressing anxieties in them. I quite understand what they mean. Death is that kind of topic. It resonates to some of our deepest fears; it pricks us to some of our stubbornest self-reassuring (but never totally convincing) rebuttals of life's tragic finiteness and our own ineffable impotence as a feather to be blown away by fate. "The wind sweeps over everyone," says the poet.

What I hope this book conveys in the end is not the tragic sense of life, but rather the place of death in life itself. The ubiquitous presence of death lends all of life—both its ecstatic and its most horrendous moments—a bitter-sweet quality. It is precisely this quality that colors so many of dualities, ambiguities, ambivalences, paradoxes, and contradictions that every sane person savors as an exciting aspect of his too brief inner existence, which we call life.

A NATIONAL SURVEY OF ATTITUDES TOWARD DEATH

Sigmund Freud said that death was the goal of all life, and Peter Pan saw it as an awfully big adventure. Socrates thought it might be the greatest of all human blessings. Readers of *Psychology Today* apparently feel that it is more important than sex; the single biggest surprise in the result of the *P.T.* death questionnaire was the sheer volume of response. More than 30,000 readers returned the research questionnaires, and more than 2,000 of them sent substantial letters with their replies. This broke the record set by the *P.T.* sex questionnaire, which fetched somewhat over 20,000 replies.

It was almost as though thousands of persons had been waiting for a legitimate occasion to unburden themselves about death and then felt somehow cleansed after writing their unspoken thoughts. Several letters said as much, indicating how grateful the respondents were and how meaningful the exercise had been to them.

These letters, clippings, quotations, and such—multipaged autobiographical accounts and philosophical treatises—helped to flush out the inescapable two-dimensional quality of the questionnaire

*From Edwin S. Shneidman, "You and Death," *Psychology Today*, June 1971. Copyright © 1971 by Communications/Research/Machines, Inc. and reproduced with their kind permission.

"You and Death" (*P.T.*, August 1970). In these writings one could see the key forces of life at work: the unconscious facets of the mind, luminescent between the lines of often poignant text.

SPECTRUM. There were seventy-five items in the questionnaire. The first section explored childhood experience of and attitudes toward death, including first-person involvement and family discussions, then influences—such as books, religion, or actual events—that might have affected early attitudes and current ones. Other questions dealt with beliefs and wishes about afterlife and with thoughts about one's own death (including the aspect that seemed most distasteful), and with feelings about the disposition of one's body. Several questions probed the problem of suicide, asking about past tries and probability of future attempts. Wills, funerals, and other death rituals also came under scrutiny.

Like other *P.T.* surveys—on sex, on law and crime, on drugs, on the cities—"Death and You" reflects the *Psychology Today* audience rather than the general American public—and in some ways it does not even totally reflect that group. While, according to recent subscriber surveys, 53 percent of the subscribers are women, women contributed 63 percent of all replies. Students also answered out of proportion to their numbers: 33 percent of all replies came from students, who make up 17 percent of the *P.T.* readership.

PROFILE. The typical respondent to the death questionnaire is a twenty- to twenty-four-year-old single, Caucasian Protestant ("somewhat religious"), politically independent ("somewhat liberal") female. She has had some college education, earns between $10,000 and $15,000 a year, lives in the Midwest, and comes from a small family (one sibling). She is in "very good" (but not "excellent") physical and mental health, and she states that there is little probability of her committing suicide.

The reader might keep in mind that responses from a young and highly educated group of people—perhaps a critical minority, a cutting edge—may reveal where the rest of society is heading.

URGENCY. The size of the response is of more than passing interest. Both the editors of *Psychology Today* and I believe that it is more than simply the habituation of *Psychology Today* readers to questionnaires as such. We believe that this volume reflects an urgency to talk about death—not only the permissiveness to talk about it. Indeed, the major debate in this country swirls around the algebra

TABLE 1
Profile of participants in *Psychology Today* death questionnaire

Sex:	
Male	37%
Female	63%
Age:	
Under 20	16%
20-24	36%
25-34	27%
35-49	15%
Fifty and over	6%
Marital status:	
Single	53%
Married, living with someone	39%
Separated, divorced, widowed	9%
Income:	
Less than $5,000	5%
$5,000-$10,000	27%
$10,000-$15,000	32%
$15,000-$25,000	25%
Over 25,000	11%
Home region:	
New England and Middle Atlantic	29%
Midwest	33%
West	22%
South	9%
Southwest and mountain states	7%
Completed level of education:	
Grade school	1%
High school	13%
Some college	38%
College graduate	17%
Some graduate work	12%
Masters degree	15%
Ph.D., M.D., other advanced degree	4%
Religious background:	
Protestant	50%
Roman Catholic	30%
Jewish	11%
Other	10%
Political attitudes:	
Very liberal	27%
Somewhat liberal	45%
Moderate	20%
Somewhat conservative	7%
Very conservative	1%

of death: Do we save more lives in a hypothetical long run by killing x number of persons today?

For many, the major result of the death questionnaire likely was that the respondent could remove death from his list of taboo topics. Death is something that most decent persons do not know how to discuss, especially with an aged or dying person. The introduction of death into such conversations usually results in embarrassment, evasion, or pretense. Along with our urgent need to discuss death, we apparently have an equally strong counterphobic reaction to it. Taken together, attraction and repulsion constitute a basic ambivalent theme that threads its way through the questionnaire. Some of our readers are quite aware of their own ambivalence.

"Reflecting upon the way I live," wrote one male student, "I know that there are things I do that make me take chances with death. The way I cross the street, the way I drive a car, the way I ski too fast, the way I sleep too little. I realize also that much of the way is in fact partly an acting out of my ambivalent feelings about life." So forbidden has death been in our culture that a third of the respondents could not recall from childhood a single instance of discussion of death within the family circle. In more than one-third of the families, it was mentioned with discomfort, and in only 30 percent was death talked about openly.

TABLE 2

When you were a child, how was death talked about in your family?

Openly	30%
With some sense of discomfort	20%
Only when necessary and then with an attempt to exclude the children	15%
As though it were a taboo subject	2%
Never recall any discussion	33%

SPLIT. Indeed, America's current attitude toward death is deeply ambivalent: awe of death and an attraction toward death; risking death and loving life; wanting happiness and behaving in self-destructive ways; regarding death as taboo and insisting on a new permissiveness to talk about it; an obsession with the Bomb and a deep concern with spiritual rebirth. We live in a death-conscious time in which man, the center of his own world, boldly asserts that he is not psychologically degradable.

Man's view of death has undergone radical change in the last few generations—perhaps the first major radical changes since the seventeenth century of Descartes. The Cartesian view of death is tied to a view of man as essentially a biological vessel, subject to the whims of fate or fortune. Death is one such whim. The Cartesian philosophic spirit necessarily implies a fatalistic view of life.

The results of the death questionnaire unequivocally demonstrate the demise of fatalism. Our reader—and perhaps most of twentieth-century men—has made himself the center of his own universe and has put himself back into his own death. He recognizes death and dying as aspects of living. Thanks primarily to Freud, he sees man as playing conscious and unconscious roles in his own fate. Most *P.T.* readers believe in the possible influence of psychological factors on man's death; 92 percent either "firmly believe" or "tend to believe" that psychological factors can influence or even cause death. But while they opt strongly for the possibility of mind over matter, half of the respondents believe that events over which individuals have no control cause most deaths. A sizable group (43 percent) believe that most persons participate consciously or unconsciously in their own deaths. Men and women generally agree in these beliefs.

TABLE 3

To what extent do you believe that psychological factors can influence (or even cause) death?

I firmly believe that they can	56%
I tend to believe that they can	36%
I am undecided or don't know	7%
I doubt that they can	2%

What is your belief about the causes of most deaths?

Most deaths result directly from the conscious efforts by the persons who die	1%
Most deaths have strong components of conscious or unconscious participation by the persons who die	43%
Most deaths just happen; they are caused by events over which individuals have no control	51%
Other	5%

WITNESS. Many things go into shaping a person's view of life and death and the part he plays in either by his own volition. The usual experiences with death in America have changed dramatically over the last two generations. It used to be that almost everyone, by the time he was an adolescent, had personally witnessed a death, usually at home, of some loved one—a baby brother or sister, a mother or father. Today most dying is done in hospitals, largely out of sight and almost always under formal institutional regimen.

Many respondents (43 percent) had their first personal involvements with death when their grandparents died. Over half first became aware of death after they were five but before they were ten, and 35 percent say that their present attitudes toward death came about from the interior—primarily from introspection and meditation.

TABLE 4

Who died in your first personal involvement with death?

Grandparents or great-grandparents	43%
Parents	8%
Brother or sister	3%
Other family member	12%
Friend or acquaintance	11%
Stranger	2%
Public figure	3%
Animal	18%

To the best of your memory, at what age were you first aware of death?

Under three	4%
Three to five	33%
Five to ten	52%
Ten or older	11%

FAITH. The interface between religion and death is especially fascinating. Numerous studies have shown that nominal religious affiliation is not meaningful in understanding religious conviction. But we asked for "religious background," which allowed us to see that one's religious environment, whether he accepts or rejects the tenets of his faith, whether he stays with the church of his fathers or leaves it, certainly affects many of his attitudes toward death. Ten percent of the respondents checked "other" rather than Protestant, Catholic, or Jewish to describe their religious backgrounds. The "other" category runs the gamut from atheism to Islam, and must include Buddhists, Hindus, Confucians, Unitarians, and complete absence of religion in the respondent's home.

Important in most current attitudes toward death is the reader's own estimate of his degree of religiosity. The "very religious" slightly outnumber the "antireligious" in our sample; roughly one out of ten *P.T.* readers falls into each group. Almost a fourth of the respondents are "not at all religious," while over half see themselves as "slightly" or "somewhat" religious. Women are more likely to be religious than men.

TABLE 5

How religious do you consider yourself to be?

Very religious	11%
Somewhat religious	32%
Slightly religious	23%
Not at all religious	24%
Antireligious	9%

AXIS. But the 10 percent of our readers who had neither Christian nor Jewish backgrounds are not the same 10 percent who are antireligious. Those who checked "other" distribute themselves over the religious spectrum, and are as likely to be "very religious" as those with Christian backgrounds. The "others" are, however, more likely (43 percent) to be "not at all religious" than are Christians, of whom about 20 percent say they are "not at all religious."

An important dimension on all *Psychology Today* questionnaires has been the respondents' self-rating on the conservative-liberal scale. *P.T.* readers rate themselves accurately; their answers to attitude questions that generally divide conservatives from liberals agree with their self-appraisals. Male readers of *P.T.* regard themselves as farther to the left than female readers do. Thirty-seven percent of the men rated themselves as "very liberal," while only 23 percent of the women placed themselves in this category.

Readers who rate themselves as "antireligious" or "not at all religious" are more likely to consider themselves "very liberal." While 27 percent of the total sample see themselves as "very liberal," 46 percent of the "antireligious" and 36 percent of the respondents who are "not at all religious" consider themselves very liberal. A Christian background does not seem to provide the optimum atmosphere for the production of very liberal persons. Less than one-fourth of Protestants and Roman Catholics consider themselves among the most liberal, but 48 percent of Jews and 35 percent of "others" place themselves in that category.

HELL. Religious background makes a difference on a number of attitudes. Jews are less likely to be religious than are Christians—53 percent of respondents with Jewish backgrounds are either "anti-religious" or "not at all religious," while only 2 percent are "very religious." Jews are also the least likely of any group to have believed in heaven and hell when they were very young. Eleven percent of those with Jewish backgrounds believed in heaven-hell concepts, while 44 percent of Protestants and 73 percent of Roman Catholics believed in them.

As adults, 18 percent of Jews "tend to believe" or "strongly believe" in afterlife, and 65 percent of them are either convinced that there is no afterlife or "tend to doubt it." Forty-two percent of Protestants and 55 percent of Roman Catholics, on the other hand, either believe or tend to believe in a hereafter, while 38 percent of Protestants and 25 percent of Roman Catholics doubt or do not believe in the existence of life after death. In every category, there were more persons who wished there were an afterlife than there were who believed or tended to believe in one. Generally, the more religious one is, the more likely he is to wish for an afterlife, but 9 percent of the "very religious" and 36 percent of those whose attitudes toward death were significantly shaped by religion prefer that there be no life after death.

IMAGE. Relatively few of the respondents believe in reincarnation. Only 6 percent strongly believe in it; 12 percent tend to believe in it; 17 percent are uncertain; and 65 percent either tend to doubt it or are convinced that it cannot occur. Most persons (68 percent) are interested in having their images survive them after death—through their children, through books, or through good works. Only 10 percent said that they were totally uninterested in their postselves.

SUICIDE. In terms of religious affiliation it is interesting to keep in mind that 50 percent of all respondents were Protestants and they made 53 percent of the suicide attempts; that 30 percent were Catholics, and Catholics made 30 percent of the suicide attempts; that 11 percent were Jewish, and Jews made 6 percent of the suicide attempts; and that 10 percent were "other," and that "others" made

TABLE 6

Which of the following best describes your childhood conceptions of death?

Heaven-and-hell concept	48%
Afterlife	9%
Death as sleep	10%
Cessation of all physical and mental activity	13%
Mysterious and unknowable	10%
Other than the above	2%
No conception	4%
Can't remember	4%

To what extent do you believe in a life after death?

Strongly believe in it	23%
Tend to believe in it	20%
Uncertain	19%
Tend to doubt it	22%
Convinced it does not exist	16%

Regardless of your belief about life after death, what is your wish about it?

I strongly wish there were a life after death	55%
I am indifferent	34%
I definitely prefer that there not be a life after death	11%

11 percent of the suicide attempts. In general, these data (except perhaps for those of Jews) are consistent with nominal religious affiliation. One implication is that being Catholic is no insurance against attempting or committing suicide. At moments of narrowed vision and heightened emotion a formal belief in the sinfulness of suicide apparently is not a sufficient deterrent to the act.

VIEW. As one would expect, religion played a very significant role in the attitudes on death of an overwhelming majority of the "very religious" (77 percent). While most of the antireligious say that religion played little part in their own attitudes, a substantial number report that religion indeed had significant roles for them, apparently pushing them away from the traditional religious outlooks on death.

Readers who consider themselves "not at all religious"—i.e., nonbelievers but not hostile to organized religion—are less likely than the antireligious to attribute a very significant role to religion.

Both Roman Catholics and Protestants report that religion played a greater role in their attitudes toward death than did other groups. Fifty-three percent of the Catholics and 44 percent of Protestants say that religion played a "very significant" or a "rather significant" role in their attitudes toward death. Only 16 percent of the Jews agree; 71 percent say that it played "no role at all" or only a "minor role." Among those who never belonged to one of the major U.S. religious groups, 24 percent say that the role of religion in their own attitudes was significant, while 61 percent say its role was nonexistent or only minor.

TABLE 7

How much of a role has religion played in the development of your attitude toward death?

	No role	Rather significant	Very significant
Very religious	2%	14%	77%
Not at all religious	25%	7%	6%
Antireligious	39%	4%	15%

TABLE 8

How much of a role has religion played in the development of your attitude toward death? (all respondents)

A very significant role	22%
A rather significant role	20%
Somewhat influential, but not a major role	23%
A relatively minor role	23%
No role at all	12%

INFLUENCE. If religion did not play a significant role in forming attitudes toward death, what, in this secular age, did? Introspection and meditation, say over a third of the respondents. Given another choice and asked to apply the influences to their own deaths, more than a third reinforced their first choice by selecting existential philosophy.

TABLE 9

Which of the following most influenced your present attitudes toward death?

Death of someone close	19%
Specific reading	11%
Religious upbringing	15%
Introspection and meditation	35%
Ritual (e.g., funerals)	5%
TV, radio, or motion pictures	3%
Longevity of my family	3%
My health or physical condition	2%
Other	8%

Men are more likely to be existentialists than women are: 38 percent of male respondents chose existential philosophy, while only 28 percent of the women cast their lots with Sartre or Kierkegaard. The existentialists are as likely to be religious as they are to be atheists.

BOOKS. Asked what books or authors had had the most effect on their attitudes toward death, 41 percent of our respondents reply that no books or authors had influenced them. (Thirty-five percent of our existential respondents are nonreaders. But before McLuhanites take too much satisfaction in these statistics, they should consider that while 11 percent say that reading had been the most important influence on their attitudes, only 3 percent credit TV, radio, or motion pictures.)

After the Bible, which influenced 23 percent, the most influential works were those of Albert Camus, Herman Hesse, and Shakespeare. James Agee influenced only 1 percent, as did Thomas Mann. Eighteen percent specify books or authors not listed on the questionnaire, and 20 percent of the existentialists are among those influenced by other writers. Among twenty-year-olds, Camus' *The Stranger* and Hesse's *Steppenwolf*, *Demian* and *Siddhartha* are mentioned again and again. Nearly two-thirds of the Camus and Hesse readers also chose existential philosophy as most influential.

Existential despair does not weight heavily on our existentialists. They report themselves no more likely than other respondents to attempt suicide; 81 percent of them feel good, cheerful, wonderful, or on top of the world; they are more likely than the average respondent to consider death as the final process of life; they are more likely than the other respondents to feel resolved when they consider their own mortality.

BOMB. Since 1945 man has faced the possibility of destruction on a scale that he has never known before: the Bomb. More than half the respondents have been influenced to some degree by the threat of nuclear war. One young man put his fear into words: "I am part of the nuclear generation. The hippie's interest in Eastern religions and emphasis on self rather than social change is in the same vein as this persistent threat of death. Philosophies are undoubtedly influenced by the atomic age and I am a product of that age."

TABLE 10

To what extent has the possibility of massive human destruction by nuclear war influenced your present attitudes toward death or life?

Enormously	5%
To a fairly large extent	14%
Moderately	18%
Somewhat	15%
Very little	25%
Not at all	23%

DRUGS. Fifteen percent of all 20- to 24-year-olds who returned the questionnaire report that narcotic or hallucinogenic drugs have affected their attitudes toward death; thus one out of every seven persons in this age group has had a significant death-and-drug experience. Drugs appear to be confined mainly to the young; only scattered returns came in from drug users over thirty-five. When all respondents are taken into consideration, men are more likely to have taken drugs than women are. Forty percent of men and 31 percent of the women have used narcotic or hallucinogenic drugs.

Religious background affects one's use of drugs. The ghostly fingers of puritanism appear to restrain Protestants, but only to a slight degree; 32 percent of them have taken drugs, and only 5 percent have been affected by their use. Among Roman Catholics, 34 percent have taken drugs, but 11 percent have had drugs affect their attitudes toward death. More Jews (44 percent) than non-Jews have tried drugs, but fewer have been affected. Only 4 percent have experienced changes in attitudes toward death after drug use. "Others" are more susceptible: 41 percent have taken drugs, but 18 percent of this group—one out of every six—say that drugs have affected their attitudes toward death. Religiosity also helps predict drug use. The more religious one says he is, the less likely he is to have taken drugs.

TABLE 11

Have your attitudes toward death ever been affected by narcotic or hallucinogenic drugs?

	Under 20	20-24	25-29	30-34
Yes	9%	15%	4%	4%
I have taken drugs but my attitudes toward death have never been affected by them	28%	34%	25%	24%
I have never taken drugs	63%	52%	71%	72%

AGE. If we put together attitudes at various ages in reference to beliefs, we find—with no great surprise—that questionnaire results show that attitudes toward death change as one matures. Typically, religious beliefs become phenomenological or secular or scientized. One then sees death simply as the end of life. The typical childhood conception of death is in terms of an afterlife, which for most involves ideas of heaven and hell (57 percent). But by adulthood,

TABLE 12

What does death mean to you?

The end; the final process of life	35%
The beginning of a life after death; a transition, a new beginning	13%
A joining of the spirit with a universal cosmic consciousness	12%
A kind of endless sleep; rest and peace	9%
Termination of this life but with survival of the spirit	17%
Don't know	10%
Other	4%

the percentage of individuals who believe in an afterlife as their primary view of death has been cut almost in half, to 30 percent.

From late adolescence on, the largest single percentage group sees death simply as the final process of life (35 percent). Among adults, beliefs about death seem to be in terms of eternal loss of consciousness—the absolute end of one's mental (including spiritual) existence. In this belief, there are no remarkable sex differences, but the older one is, the more likely he is to be convinced that there is no life after death.

MEANING. Consistent with this view, the most distasteful aspect of death is that "one can no longer have any experiences" (36 percent). Again we found traces of the Protestant ethic at work. Fifty-six percent of Jewish respondents agree that the inability to have experiences is the most distasteful aspect of their own deaths. Almost no Jews or Roman Catholics (1 percent) said they were most concerned about their dependents, but 7 percent of the Protestants find that their responsibilities to their families weigh heaviest of all aspects surrounding their own deaths. Respondents under twenty have not yet shaken the heaven-and-hell concept; 28 percent of them—or more than twice the average—find their fate in a possible afterworld the most distasteful aspect of their own deaths.

TABLE 13

What aspect of your own death is the most distasteful to you?

I could no longer have any experiences	36%
I am afraid of what might happen to my body after death	2%
I am uncertain as to what might happen to me if there is a life after death	12%
I could no longer provide for my dependents	4%
It would cause grief to my relatives and friends	10%
All my plans and projects would come to an end	11%
The process of dying might be painful	15%
Other	9%

WHEN. Most people believe that life's experiences should be both long in duration and wide in richness—but not at all costs. For example, most respondents believe that only those efforts that are "reasonable" for a person's age, physical and mental condition, and state of pain should be made to keep him alive (58 percent). Less than one in ten opt for "all possible effort" to keep an individual alive in any circumstances. Men are more likely than women to agree that life is sweet, even for the pain-racked and the senile.

THOUGHT. All these results point to one of the main findings of the questionnaire, namely that—over the past generation or two—there has been a tremendous secularization of death. Nowadays people die ascetically in aseptic hospitals rather than aesthetically in their homes. The physician has replaced the priest; the doctor is today's magician who has the power to extend life, our new escort from this vale of tears. The funeral industry directs the molds of mourning, ushering us from burial to bereavement.

The average respondent thinks about his own death "occasionally," and the thought is likely to make him feel resolved in relation to life or to take pleasure in being alive.

TABLE 14

How often do you think about your own death?

Very frequently (at least once a day)	5%
Frequently	17%
Occasionally	57%
Rarely (no more than once a year)	15%
Very rarely or never	6%

When you think of your own death (or when circumstances make you realize your own mortality), how do you feel?

Fearful	19%
Discouraged	5%
Depressed	11%
Purposeless	5%
Resolved in relation to life	26%
Pleasure in being alive	25%
Other	9%

There are no significant sex differences, although men are slightly more likely to feel pleasure in being alive or to be resolved, while women more often tend to be fearful. There seem to be no particular age trends in regard to this feeling, except that members of the 25-29 age group are more fearful than any other group; 30 percent say that fear is their response to thoughts of death. Perhaps it is at about this age that a person first begins to recognize his own mortality. But when respondents were asked at what age they thought other people were most afraid of death, their answers spread across the spectrum. Women, however, are a bit more likely than men to say that children under twelve are most afraid, perhaps because of their closer contact with young children. Both men and women place the time of least fear in the years over seventy, but women are almost as likely to say that the thirties are least haunted by the specter of death.

FEAR. As one grows older, he begins to feel that those younger than he are more fearful at the thought of death. Among those in their forties, for example, 25 percent say that children under twelve are most afraid of death; 14 percent of the entire sample agree with them. Thirty-eight percent of readers in their fifties say that adolescents are most afraid, while only 11 percent of the entire sample concur.

TABLE 15

In your opinion, at what age are people most afraid of death?

Up to 12 years	14%
13-19 years	11%
20-29 years	13%
30-39 years	10%
40-49 years	16%
50-59 years	16%
60-69 years	13%
70 years and over	8%

TIME. Almost no one wishes to die in youth or in the middle prime of life. Two-thirds of the respondents would like to die in old age, but fewer people would like to live to ripe age than believe that they will. More men (74 percent) than women (62 percent) want to live out their full life spans.

GREETING. By and large, *P.T.* readers are either death-postponers (39 percent) or death-accepters (42 percent). Fourteen percent are death-fearers, while 2 percent say they are death-welcomers, and 1 percent falls into each of the remaining categories, death-seekers

TABLE 16

If you could choose, when would you die?

In youth	2%
In the middle prime of life	3%
Just after the prime of life	29%
In old age	66%

When do you believe that, in fact, you will die?

In youth	2%
In the middle prime of life	10%
Just after the prime of life	19%
In old age	69%

and death-hasteners. There are no remarkable sex differences, but all death-hasteners feel either "down and out" or "kind of low," while 100 percent of respondents who feel "on top of the world" and 88 percent of those who feel "wonderful" are either postponers or accepters. Oddly enough, over half of the death-welcomers feel "on the whole, all right," "cheerful," or "wonderful."

Respondents would prefer quiet and dignified deaths, sudden but not violent. Eight out of 10 would not, even if it were possible, want to know the exact date on which they are going to die, but if they had a terminal disease, seven out of ten would want their physicians to tell them so. Men are more likely than women to want to know the exact date of their deaths.

TABLE 17

If you had a choice, what kind of death would you prefer?

Tragic, violent death	1%
Sudden, but not violent death	38%
Quiet, dignified death	30%
Death in line of duty	1%
Death after a great achievement	6%
Suicide	2%
Homicidal victim	0%
There is no "appropriate" kind of death	16%
Other	6%

Further, if they had a terminal disease and limited time to live, nearly half would either shift their concerns to others or finish projects and tie up loose ends. Even though over a third formed their basic attitudes toward death through contemplation and introspection, only one in twenty would spend his remaining days contemplating or praying. Twenty percent would make no change in life style, and 19 percent would satisfy hedonistic needs through travel, sex, or drugs.

FACING. Perhaps because so many men have seen combat in this war-torn century, more men than women have been in situations in which they seriously thought they might die. Seventy-nine percent of the men and 70 percent of the women have faced death at least once, and 28 percent of the men as compared to 20 percent of the women have faced death "several" or "many" times.

Only 16 percent hold life so dear that they would not sacrifice it for any reason. More men than women report themselves willing to sacrifice their lives for an idea or a moral principle; women are more likely to say they would sacrifice their lives for loved ones. Age also affects one's attitude, respondents under twenty are least eager to sacrifice their lives for those they love, but are more willing than most other groups to die for an idea or in combat. Respondents from thirty-five to thirty-nine value their own lives more highly than members of any other age group; 30 percent would not give their lives for any reason. As Mark Twain said, "Each person is born to one possession which out-values all others—his last breath."

TABLE 18

For whom or what might you be willing to sacrifice your life?

For a loved one	58%
For an idea or a moral principle	15%
In combat or a grave emergency where a life could be saved	11%
Not for any reason	16%

TWO. Arnold Toynbee has written that death is essentially a two-person affair, involving both the survivor and the decedent. He further asserts that if a married person truly loves the spouse, that person will wish the spouse to die first, so the spouse will be spared the anguish of bereavement. Respondents to the death questionnaire divide on this issue, an equal percentage (23 percent) saying yes and no; most persons (54 percent) cannot make up their minds. But when we ignore the undecided, we find a definite split between the sexes. Among the men, 33 percent want to outlive their wives, while 18 percent of the women want to outlive their husbands. Eleven percent of the men and 30 percent of the women would like to die first. Housewives have made up their minds; only 38 percent are undecided, and 35 percent would prefer to outlive their husbands—twice as many as women readers on the average. Only the business managers and the executives are as likely to want to survive their spouses. Readers over thirty are more likely to say they wish to outlive their spouses. The reasons they give for "yes" and "no" responses are themselves equally divided between selfish and selfless reasons. One is reminded of what Lord Nelson is reported to have said as he lay dying aboard the *Victory:* "The pain is so great that one might wish oneself dead but one would like to live a little longer, too." As ambivalence is the keystone of life, so is it a characteristic of death.

TABLE 19

What is your primary reason for your wish either to predecease or to outlive your spouse?

To spare my spouse loneliness	6%
To avoid loneliness for myself	9%
To spare my spouse grief	5%
To avoid grief for myself	5%
Because the surviving spouse could cope better with grief or loneliness	10%
To live as long as possible	13%
None of the above	37%
Other	15%

READY. Well over half the respondents have wanted to die at one time or another in their lives but the sexes differ sharply. While 47 percent of the men have wished they were dead, 68 percent of the women have wished for death on some occasion—most of them because of "great emotional upset."

Although 60 percent of all who answered have wished for death and 53 percent have seriously contemplated killing themselves, only 13 percent have actually attempted suicide. Like the Kinsey data on sexual behaviors and patterns, the data from the death questionnaire on past suicide attempts are revealing: apparently 3,900 people out of 30,000 have tried to kill themselves. Among those who tried, 63 percent have made attempts unlikely to result in death; 23 percent have made attempts with moderate probability of death; and 14 percent have made attempts of high lethality. This means that around 540 of the respondents may have histories of extremely serious tries at suicide.

Almost a quarter of those who answered report that they contemplate suicide "once in a while" or "very often," and women are twice as likely as men to think frequently of killing themselves (8 percent of the women compared with 4 percent of the men). Respondents under twenty are more likely than average (11 percent as compared to 6 percent) to contemplate suicide "very often." One-third of *P.T.* readers are sure that they would never commit suicide; another 41 percent doubt that there is any possibility that they would; and 81 percent are sure that they will not commit suicide in the near future. But three percent are afraid that they might kill themselves and 1 percent plan to commit suicide someday.

TABLE 20

Has there been a time in your life when you wanted to die?

Yes, mainly because of great physical pain	2%
Yes, mainly because of great emotional upset	37%
Yes, mainly to escape an intolerable social or interpersonal situation	18%
Yes, mainly because of great embarrassment	1%
Yes, for a reason other than above	3%
No	40%

MOTIVE. What might drive a *P.T.* reader to suicide? Loneliness, illness, or physical pain, say 55 percent. Only 2 percent say they would kill themselves to get even or hurt someone. Women are more likely than men to give loneliness, the loss of a loved one, or atomic war as a motive for suicide, while men are more likely to say they might kill themselves because of failure or disgrace or to avoid sickness or physical pain. One wonders about the courage of the stronger sex when 33 percent of the male readers say that physical illness or pain could drive them to commit suicide. Only 19 percent of the women say they might kill themselves for that reason. Anticipated pain may be more frightening than actual pain;

only 2 percent of respondents say they have ever wished for death as escape from physical pain. The older one is, the more likely he is to say that illness or pain could motivate his suicide.

TABLE 21

Suppose that you were to commit suicide; what reason would most motivate you to do it?

To get even or hurt someone	2%
Fear of insanity	6%
Physical illness or pain	24%
Failure or disgrace	5%
Loneliness or abandonment	31%
Death or loss of a loved one	9%
Family strife	1%
Atomic war	8%
Other	15%

METHOD. Both men and women select pills as the preferred method for suicide, but women are even more likely than men to choose this quiet path to death. Sixty-one percent of the men and 74 percent of the women would take barbiturates or pills. There is no close second, but more men than women would shoot themselves. Fifteen percent of the men and only 4 percent of the women prefer a bullet.

More than half the respondents say that if they commit suicide, they will leave notes—a figure far in excess of what one actually finds in the files of coroners' offices: fewer than 15 percent of most suicides take the time to leave a note. While women divide almost equally on the question, two-thirds of the men say they would leave suicide notes.

Most persons (85 percent) believe that suicide should be prevented, although only 32 percent say that there are no circumstances in which a person should be allowed to take his own life. A minority (15 percent) believes that if a person wants to commit suicide, society has no right to stop him.

PROVISION. In general, *P.T.* readers are generous and "other-oriented," and—in spite of their disavowal of an interest in their postselves—they do in large numbers, believe in providing for others through insurance (83 percent) and through wills (84 percent). Only 1 percent says, "I definitely won't leave a will," and

TABLE 22

Suppose you were to commit suicide; what method would you be most likely to use?

Barbiturates or pills	69%
Gunshot	8%
Hanging	1%
Drowning	1%
Jumping	2%
Cutting or stabbing	1%
Carbon monoxide	10%
Other	7%

only 3 percent state that they definitely do not believe in insurance, that they "do not have and do not plan to get insurance."

Women, who are generally the beneficiaries of insurance policies, tend to doubt the value of insurance more frequently than men do. Eighteen percent of the women and 7 percent of the men are either "undecided" or "tend not to believe" in providing for their survivors through life insurance.

RITES. Their attitudes about funerals are quite definite and echo Jessica Mitford's *The American Way of Death.* Most (80 percent) believe that funerals are "very much overpriced," and 62 percent believe that funerals should cost less than $300. Only 2 percent consider a price in excess of $600 to be a "reasonable" charge.

Only 2 percent would like formal funerals—"as large as possible"—and a third of *P.T.* readers want no funerals of any kind. Large, formal funerals are rejected unanimously by respondents who are antireligious, who want no part of any afterlife, or who have religious backgrounds that are neither Jewish, Catholic, nor Protestant. The less religious one is, the less likely he is to want a funeral. Among the very religious, 20 percent do not want funerals; the prejudice against these rites climbs steadily through the categories of religiosity, with 63 percent of the antireligious saying they do not want funerals. Readers believe that rituals such as funerals and wakes are rather important for the survivors (47 percent), although almost a fifth (18 percent) say that such rituals are not important at all. Readers with Jewish backgrounds are more likely than members of any other group to say that whatever their survivors want is fine with them.

Almost no one (6 percent) approves of lying in state at his own funeral. One-fourth of all respondents dissociate themselves from their bodies and say they just "don't care." But a substantial majority (70 percent) definitely do not want to lie in open caskets. Women tend to feel more strongly against lying in open caskets than do men. Antireligious readers register overwhelming disapproval (91 percent) against the display of their corpses, and Roman Catholics are most likely to approve than any other group (12 percent).

BODIES. If it were entirely up to them, nearly a third (32 percent) of the respondents would donate their bodies to medical schools or to science. There is a very slight tendency among those who believe or tend to believe in reincarnation to be readier than average to donate their bodies to science. This tendency also exists among people who prefer that there be no afterlife.

Most respondents are willing to donate their hearts for transplantation; 82 percent would—after their own deaths—donate their hearts to anyone, and another 3 percent would donate their hearts to friends or relatives. While 15 percent of all readers would decline to make their hearts available, the very religious are most likely to refuse to give their hearts (25 percent).

Among readers who do not wish to donate their bodies to science, there are more who choose cremation than there are who select burial. But the antireligious are overwhelmingly against burial; only 2 percent would allow their bodies to be interred in caskets, and over half (52 percent) would like to be cremated. Religious background plays a part in one's choice. A third of all readers with Jewish or Roman Catholic backgrounds want to be buried, while only 13 percent of the Protestants would agree with them. Protestants are likelier to donate their bodies or to choose cremation than either Jews or Roman Catholics.

Most approve of autopsies; 83 percent either approve of them or have no strong feelings against having autopsies performed on their own bodies. There is a slight tendency for men to be more in favor of autopsies than for women to approve them. One's wish for an afterlife affects one's feeling toward autopsies. The less one wishes for an afterlife, the more likely he is to approve of autopsies. While

79 percent of those who strongly wish for a hereafter do not object to autopsies, 93 percent of those who want the grave to be the end have no strong feelings against them.

TABLE 23

If it were entirely up to you, how would you like to have your body disposed of after you have died?

Burial	22%
Cremation	31%
Donation to medical school or science	32%
I am indifferent	16%

Respondents to the death questionnaire furnish a few clues for our heightened preoccupation with death today. A sizable percentage (45 percent) attribute the increase in concern within the last twenty-five years to wars and to the presence of the Bomb. This is not the place to dwell on the complicated morality of the war in Indochina, but perhaps even more important is the mushroom cloud of death represented by the omnipresent threat of nuclear war. The largest single factor relating to death in the 1970s may well be the threat of atomic bombs, and in a chain reaction, it might be that this heightened concern with atomic deaths accounts in part for our outwardly disdainful attitude toward life itself: excessive risk-taking, burning out with drugs, daring authority, flaunting tradition, and the many other inimical and destructive forms of behavior toward ourselves and toward our institutions and traditions. One young college woman wrote: "I can't say that I think this or do such and such because we all may die tomorrow, but I'm convinced that in all our minds there's an underlying nagging fear that the world could blow up, a fear which today is more silent than in the bomb-shelter era of a few years ago but perhaps even more strongly felt and exhibited in minute, imperceptible ways."

But we human beings, especially the young ones, fight on. Another twenty-year-old wrote about this questionnaire: "If I pass this test do I get immortal life as a prize? Think what a bummer that would be. Fear of death puts a little excitement into life. R.I.P."

BIBLIOGRAPHY

Abbott, Samuel W. 1901. "Death Certification," in *Reference Handbook of the Medical Sciences,* ed. Albert H. Buck. New York: William Wood.

Agee, James. 1959. *A Death in the Family.* New York: Avon.

Anders, Gunther. 1962. "Reflections on the H Bomb," in *Man Alone: Alienation in Modern Society,* ed. Eric and Mary Josephson. New York: Dell.

Bailey, Richard M. 1970. "Economic and Social Costs of Death," in *The Dying Patient,* ed. Orville G. Brim, Jr., et al. New York: Russell Sage Foundation.

Bailyn, Bernard. 1967. *Ideological Origins of the American Revolution.* Cambridge: Belknap Press, Harvard University Press.

Beauvoir, Simone de. 1966. *A Very Easy Death.* New York: Putnam.

Beecher, Henry K., et al. 1968. "A Definition of Irreversible Coma." *Journal of the American Medical Association* 205:85–88.

Black, Henry Campbell. 1968. *Black's Law Dictionary,* rev. ed. St. Paul, Minn.: West Publishing Co.

Bridgman, Percy W. 1938. *The Intelligent Individual and Society.* New York: Macmillan.

Brim, Orville G., Jr., et al., eds. 1970. *The Dying Patient.* New York: Russell Sage Foundation.

Caillois, Roger. 1961. *Man, Play, and Games.* New York: Free Press, Macmillan.

Camus, Albert. 1955. *Myth of Sisyphus and Other Essays*. New York: Knopf.

Caplan, Gerald. 1964. *Principles of Preventive Psychiatry*. New York: Basic Books.

Cassedy, James H. 1961. *Demography in Early America*. Cambridge: Harvard University Press.

Choron, Jacques. 1963. *Death and Western Thought*. New York: Collier Books.

Colby, Marian. 1965. "The Significance, Evolution, and Implementation of Standard Certificates." *American Journal of Public Health*, 55: 596–99.

Crane, Diana. 1970. "Dying and Its Dilemmas as a Field of Research," in *The Dying Patient*, ed. Orville G. Brim, Jr., et al. New York: Russell Sage Foundation.

Crane, Stephen. 1967. *The Red Badge of Courage*. New York: Macmillan. Originally published 1895.

Curphey, Theodore J. 1961. "The Role of the Social Scientist in the Medicolegal Certification of Death from Suicide," in *The Cry for Help*, ed. Norman L. Farberow and Edwin S. Shneidman. New York: McGraw-Hill.

———. 1967. "The Forensic Pathologist and the Multi-Disciplinary Approach to Death," in *Essays in Self-Destruction*, ed. Edwin S. Shneidman. New York: Science House.

Davis, Merrell R., and William H. Gilman, eds. 1960. *Letters of Herman Melville*. New Haven: Yale University Press.

Denniston, Robin. 1969. Foreword to *Man's Concern with Death*, ed. Arnold Toynbee et al. New York: McGraw-Hill.

Draper, G., C. W. Dupertuis, and J. L. Caughley. 1944. *Human Constitution in Clinical Medicine*. New York: Hoeber.

Ducasse, C. J. 1961. *The Belief in a Life After Death*. Springfield, Ill.: Charles C. Thomas.

Eissler, Kurt. 1955. "Death and the Pleasure Principle," in *Psychiatry and the Dying Patient*. New York: International Universities Press.

Elliot, Gil. 1972. *Twentieth Century Book of the Dead*. New York: Charles Scribner's Sons.

Erikson, Erik. 1963. *Childhood and Society*, 2nd ed. New York: Norton.

Farber, Maurice L. 1968. *Theory of Suicide*. New York: Funk & Wagnall.

Feifel, Herman. 1963. "Death," in *Taboo Topics*, ed. Norman L. Farberow. New York: Atherton Press.

———. 1965. "The Function of Attitudes Toward Death," in *Death and Dying: Attitudes of Patient and Doctor*. New York: Group for the Advancement of Psychiatry.

———, ed. 1959. *The Meaning of Death.* New York: McGraw-Hill.

Forbes, Thomas R. 1970. "Life and Death in Shakespeare's London." *American Scientist* 58:511–20.

Freidman, Paul, and L. Lum. 1957. "Some Psychiatric Notes on the *Andrea Doria* Disaster." *American Journal of Psychiatry* 114:426-32.

Freud, Sigmund. 1915. "Thoughts for the Times on War and Death," in *Standard Edition of the Complete Works.* London: Hogarth Press.

Fulton, Robert, ed. 1965. *Death and Identity.* New York: Wiley.

Glaser, Barney G., and Anselm L. Strauss. 1965. *Awareness of Dying.* Chicago: Aldine.

———. 1968. *Time for Dying.* Chicago: Aldine.

Glaser, Robert J. 1970. "Innovations and Heroic Acts in Prolonging Life," in *The Dying Patient,* ed. Orville G. Brim, Jr., et al. New York: Russell Sage Foundation.

Gorer, Geoffrey. 1965. *Death, Grief, and Mourning.* Garden City, N. Y.: Doubleday.

Graunt, John. *Natural and Political Annotations . . . upon the Bills of Mortality.* London, 1662.

Gross, Charles. 1896. "Select Cases from the Coroners' Rolls A.D. 1265–1413, with a Brief Account of the History of the Office of Coroner." London: Bernard Quaritch.

Haydn, Hiram. 1969. "Portrait: Henry A. Murray." *American Scholar,* 39:123-136.

Henderson, Joseph L., and Maud Oakes. 1963. *The Wisdom of the Serpent: The Myths of Death, Rebirth, and Resurrection.* New York: Braziller.

Herzog, Alfred. 1968. "A Clinical Study of Parental Response to Adolescent Death by Suicide with Recommendations for Approaching the Survivors," in *Proceedings of the Fourth International Conference for Suicide Prevention,* ed. Norman L. Farberow. Los Angeles: Delmar.

Hesse, Hermann. 1969. *Narcissus and Goldmund,* trans. Ursule Molinaro. New York: Noonday Press.

———. 1970. *Klingsor's Last Summer,* trans. Richard and Clara Winston. New York: Noonday Press.

Hinton, John. 1967. *Dying.* Baltimore: Penguin Books.

Hotchner, A. E. 1966. *Papa Hemingway: A Personal Memoir.* New York: Random House.

Huizinga, Johan. 1950. *Homo Ludens: A Study of the Play Element in Culture.* Boston: Beacon Press. Originally published 1944.

Hunnisett, R. F. 1961. *The Medieval Coroner.* Cambridge: Cambridge University Press.

James, William. 1890. *Principles of Psychology*. New York: Henry Holt.

Kahn, Herman. 1960. *On Thermonuclear War*. Princeton: Princeton University Press.

Kalish, Richard. 1965. "Some Variables in Death Attitudes," in *Death and Identity*, ed. Robert Fulton. New York: Wiley.

———. 1968. "Life and Death: Dividing the Indivisible." *Social Science and Medicine* 2:249–59.

Kaplan, Abraham. 1970. *The Self and Identity*. Unpublished. University of Haifa.

Kargon, Robert. 1963. "John Graunt, Francis Bacon, and the Royal Society: The Reception of Statistics." *Journal of the History of Medicine* 18:337–48.

Kastenbaum, Robert, and Ruth Aisenberg. 1972. *The Psychology of Death*. New York: Springer.

Keniston, Kenneth. 1968. *Young Radicals*. New York: Harcourt, Brace & World.

Kleitman, Nathaniel. 1963. *Sleep and Wakefulness*, rev. ed. Chicago: University of Chicago Press.

Knutson, Andie L. 1970. "Cultural Beliefs on Life and Death," in *The Dying Patient*, ed. Orville G. Brim, Jr., et al. New York: Russell Sage Foundation.

Kobler, Arthur, and Ezra Stotland. 1964. *The End of Hope*. New York: Free Press of Glencoe.

Kübler-Ross, Elisabeth. 1969. *On Death and Dying*. New York: Macmillan.

———. 1972. Review of Hinton's *Dying*. *Life-Threatening Behavior* 2:56-58.

Laing, R. D. 1967. *The Politics of Experience*. New York: Pantheon Books.

Lamont, Corliss. 1950. *The Illusion of Immortality*, 2nd ed. New York: Philosophical Library.

Langer, Suzanne. 1967. *Mind: An Essay in Human Feeling*. Baltimore: Johns Hopkins Press.

Lifton, Robert Jay. 1967. *Death in Life: Survivors of Hiroshima*. New York: Random House.

Lindemann, Erich. 1944. "Symptomatology and Management of Acute Grief." *American Journal of Psychiatry* 101:141–48.

———. 1964. "Immobilization Response to Suicidal Behavior." *Archives of General Psychiatry*. 11:282-285.

Litman, Robert E. 1967. "Sigmund Freud on Suicide," *in Essays in Self-Destruction*, ed. E. S. Shneidman. New York: Science House.

———. 1970. "Medical-Legal Aspects of Suicide," in *The Psychology of Suicide,* ed. E. S. Shneidman, N. L. Farberow, and R. E. Litman. New York: Science House.

———, T. J. Curphey, E. S. Shneidman, N. L. Farberow, and N. D. Tabachnick. 1963. "Investigations of Equivocal Suicides." *Journal of the American Medical Association* 184:924–29.

Logan, W. P. D. 1969. "Vital Statistics," in *Theory and Practice of Public Health,* ed. W. Hobson, 3rd ed. New York: Oxford University Press.

McClelland, David. 1963. "The Harlequin Complex," in *The Study of Lives,* ed. Robert W. White. New York: Atherton.

MacDonald, John M. 1961. *The Murderer and His Victim.* Springfield, Ill.: Charles C. Thomas.

Manning, Bayless. 1970. "Legal and Policy Issues in the Allocation of Death," in *The Dying Patient,* ed. Orville G. Brim, Jr., et al. New York: Russell Sage Foundation.

Mant, A. Keith. 1969. "The Medical Definition of Death," in *Man's Concern with Death,* ed. Arnold Toynbee et al. New York: McGraw-Hill.

Maslow, Abraham. Editorial in *Psychology Today,* August 1970.

Mitford, Jessica. 1963. *The American Way of Death.* New York: Simon & Schuster.

Montale, Eugenio. 1960. "Billy Budd." *Sewanee Review* 68: 419-22

Muggeridge, Malcolm. *The Observer,* February 20, 1970.

Murray, Henry A. 1940. "What Should Psychologists Do About Psychoanalysis?" *Journal of Abnormal and Social Psychology* 35:150–75.

———. 1949. Introduction, Melville's *Pierre.* New York: (Hendricks House), Farrar Straus. Pp. xiii–ciii.

———. 1951. "In Nomine Diaboli." *New England Quarterly,* 24: 435–52. Reprinted in *Moby-Dick Centennial Essays.* 1953. Dallas: Southern Methodist University Press; *Discussions of Moby Dick.* 1960. Boston: D. C. Heath & Co.; *Melville: A Collection of Critical Essays.* 1962. Englewood Cliffs, New Jersey: Prentice-Hall; *Psychology Today,* September, 1968.

———. 1959. "Preparations for the Scaffold of a Comprehensive System." In *Psychology: A Study of a Science,* ed. S. Koch, Vol. 3. New York: McGraw-Hill.

———. 1966. "Bartleby and I." In *Bartleby the Scrivener,* ed. H. P. Vincent. Kent, Ohio: Kent State University Press.

———. 1967. "Henry A. Murray." In *The History of Psychology in Autobiography,* eds. E. G. Boring and G. Lindzey. New York: Appleton-Century-Crofts.

———. 1957. "Morton Prince: Sketch of His Life and Work," in *Harvard Psychological Clinic, 1927-1957*. Cambridge: Harvard University Press.

———. 1967. "Dead to the World: The Passions of Herman Melville," in *Essays in Self-Destruction*, ed. E. S. Shneidman. New York: Science House.

Naegele, Kaspar D. 1970. *Health and Healing*. San Francisco: Jossey-Bass.

Parkes, C. Murray. 1970. "The First Year of Bereavement: A Longitudinal Study of the Reaction of London Widows to the Death of Their Husbands." *Psychiatry* 33:444–67.

———. 1971. "Psycho-Social Transitions: A Field for Study." *Social Science and Medicine* 5:101–15.

——— et al. 1969. "Broken Heart: A Statistical Study of Increased Mortality Among Widowers." *British Medical Journal* 1:740–43.

———. 1972. *Bereavement*. New York: International Universities Press.

Pepper, Stephen C. 1967. "Can a Philosophy Make One Philosophical?" in *Essays in Self-Destruction*, ed. Edwin S. Shneidman. New York: Science House.

Platt, John. 1969. "What We Must Do." *Science* 166:1115–21.

Riley, John W., Jr., 1970. "What People Think About Death," in *The Dying Patient*, ed. Orville G. Brim, Jr., et al. New York: Russell Sage Foundation.

Rogow, Arnold A. 1963. *James Forrestal: A Study of Personality, Politics, and Policy*. Los Angeles: Boulevard Bookshop.

Russell, Bertrand. 1967. *The Autobiography of Bertrand Russell, 1872–1914*, 2 vols. Boston: Little, Brown.

Sato, Koji. 1966. "Discussion of Orientations Toward Death." *International Journal of Psychiatry* 2:196–97.

Serrano, Miguel. 1966. *C. G. Jung and Hermann Hesse: A Record of Two Friendships*. New York: Schocken Books.

Shiga, Naoya. 1956. "Han's Crime." In *Modern Japanese Literature*, ed. Donald Keene. New York: Grove Press. (Translated by Ivan Morris).

Shneidman, Edwin S. 1963. "Orientations Toward Death: A Vital Aspect of the Study of Lives," in *The Study of Lives*, ed. Robert W. White. New York: Atherton.

———. 1967. "Sleep and Self-Destruction: A Phenomenological Study," in *Essays in Self-Destruction*, ed. E. S. Shneidman. New York: Science House.

———. 1968a. "Orientations Toward Cessation: A Re-Examination of Current Modes of Death." *Journal of Forensic Sciences* 13:33–45.

———. 1968b. "Suicide Prevention: A Current National View," in *Proceedings of the Fourth International Conference for Suicide Prevention*, ed. Norman L. Farberow. Los Angeles: Delmar.

———. 1969. "Suicide, Lethality, and the Psychological Autopsy," in *Aspects of Depression,* ed. E. S. Shneidman and M. Ortega. Boston: Little, Brown.

———. 1970a. "Suicide, Sleep, and Death: Some Possible Interrelations and Cessation, Interruption, and Continuation Phenomena," in *The Psychology of Suicide,* ed. E. S. Shneidman, N. L. Farberow, and R. E. Litman. New York: Science House. Originally published 1964.

———. 1970b. "The Deaths of Herman Melville," in *The Psychology of Suicide,* ed. E. S. Shneidman, N. L. Farberow, and R. E. Litman. New York: Science House.

———. 1971a. "You and Death." *Psychology Today,* June:43–45, 74–80.

———. 1971b. "Prevention, Intervention, and Postvention of Suicide." *Annals of Internal Medicine* 75:453–58.

———, ed. 1967. *Essays in Self-Destruction.* New York: Science House.

——— and Norman L. Farberow. 1961. "Sample Investigations of Equivocal Deaths," in *The Cry for Help,* ed. N. L. Farberow and E. S. Shneidman. New York: McGraw-Hill.

Silverman, Phyllis R. 1969. "The Widow-to-Widow Program: An Experiment in Preventive Intervention." *Mental Hygiene* 53:333–37.

Sudnow, David. 1967. *Passing On.* Englewood Cliffs, N.J.: Prentice-Hall.

———. 1970. "Dying in a Public Hospital," in *The Dying Patient,* ed. Orville G. Brim, Jr., et al. New York: Russell Sage Foundation.

Sussmilch, Johann. 1741. *Die Göttliche Ordnung in den Veränderungen des Menschlichen Geschlechts, aus der Geburt, dem Tode, und der Fortpflanzung desselben erwiesen.* Berlin: J. C. Spenser.

Toynbee, Arnold. 1969. "Why and How I Work." *Saturday Review,* April 5:22–27, 62.

——— et al., eds. 1969. *Man's Concern with Death.* New York: McGraw-Hill.

Unamuno, Miguel de. 1954. *Tragic Sense of Life.* New York: Dover Publications. Originally published 1921.

United Nations. 1955. *Handbook of Vital Statistics Methods.* New York.

U.S. Department of Health, Education and Welfare. 1965. *Mortality Trends in the United States, 1954–1963.* Washington, D.C.: U.S. Government Printing Office.

———. 1966. *Report of the United States Delegation to the International Conference for the Eighth Revision of the International Classification of Diseases, Geneva, Switzerland, July 6–12, 1965.* Washington, D.C.: U.S. Government Printing Office.

Varah, Chad. 1966. *The Samaritans.* New York: Macmillan.

Viscott, David S. 1970. "A Musical Idiot Savant: A Psychodynamic Study and Some Speculations on the Creative Process." *Psychiatry* 33: 494–515.

Vonnegut, Kurt. 1969. *Slaughterhouse Five*. New York: Dell.

Wald, George. 1969. "A Generation in Search of a Future." *The Boston Globe,* March 5.

Waldo, F. J. 1910. "The Ancient Office of Coroner." *Coroners' Society Annual Report* 4:241–52.

Wallace, A. F. 1956. *Tornado in Worcester: An Exploratory Study of Individual and Community Behavior in an Extreme Situation.* National Academy of Sciences Disaster Study No. 3. Washington, D.C.: National Research Council.

Weisman, Avery. 1966. "Discussion of Suicide and Appropriate Death." *International Journal of Psychiatry* 2:190–93.

———. 1972. *On Dying and Denying*. New York: Behavioral Publications.

——— and Thomas Hackett. 1961. "Predilection to Death." *Psychosomatic Medicine* 23:232–56.

——— and ———. 1962. "The Dying Patient." *Forest Hospital Publications* 1:16–21.

——— and Robert Kastenbaum. 1968. *The Psychological Autopsy*. Community Mental Health Journal Monograph No. 4. New York: Behavioral Publications.

Wertenbaker, Lael Tucker. 1957. *Death of a Man.* New York: Random House.

Wittgenstein, Ludwig. 1922. *Tractatus Logico-Philosophicus.* London: Routledge & Kegan Paul.

Wolfenstein, Martha. 1957. *Disaster: A Psychological Essay.* New York: Free Press, Macmillan.

Wolfgang, Marvin E. 1958. *Patterns in Criminal Homicide*. Philadelphia: University of Pennsylvania Press.

———. 1959. "Suicide by Means of Victim-Precipitated Homicide." *Journal of Clinical and Experimental Psychopathology* 20:335-49.

World Health Organization. 1957. *Manual of the International Statistical Classification of Diseases, Injuries, and Causes of Death: Based on the Recommendations of the Seventh Revision Conference, 1955*. Geneva: World Health Organization.

NAME INDEX

SUBJECT INDEX